Lecture Notes in Civil Engineering

Volume 123

Lecture Notes in Civil Engineering (LNCE) publishes the latest developments in Civil Engineering - quickly, informally and in top quality. Though original research reported in proceedings and post-proceedings represents the core of LNCE, edited volumes of exceptionally high quality and interest may also be considered for publication. Volumes published in LNCE embrace all aspects and subfields of, as well as new challenges in, Civil Engineering. Topics in the series include:

- Construction and Structural Mechanics
- Building Materials
- Concrete, Steel and Timber Structures
- Geotechnical Engineering
- Earthquake Engineering
- Coastal Engineering
- Ocean and Offshore Engineering; Ships and Floating Structures
- Hydraulics, Hydrology and Water Resources Engineering
- Environmental Engineering and Sustainability
- Structural Health and Monitoring
- Surveying and Geographical Information Systems
- Indoor Environments
- Transportation and Traffic
- Risk Analysis
- Safety and Security

To submit a proposal or request further information, please contact the appropriate Springer Editor:

- Mr. Pierpaolo Riva at pierpaolo.riva@springer.com (Europe and Americas);
- Ms. Swati Meherishi at swati.meherishi@springer.com (Asia - except China, and Australia, New Zealand);
- Dr. Mengchu Huang at mengchu.huang@springer.com (China).

All books in the series now indexed by Scopus and EI Compendex database!

More information about this series at http://www.springer.com/series/15087

Ankit Garg · C. H. Solanki ·
Chandra Bogireddy · Junwei Liu
Editors

Proceedings of the 1st Indo-China Research Series in Geotechnical and Geoenvironmental Engineering

Springer

Editors
Ankit Garg
Guangdong Engineering Center for Structure Safety and Health Monitoring
Department of Civil and Environmental Engineering
Shantou University
Shantou, China

Chandra Bogireddy
Department of Civil Engineering
Vardhaman College of Engineering
Hyderabad, India

C. H. Solanki
Department of Civil Engineering
Sardar Vallabhbhai National Institute of Technology
Surat, India

Junwei Liu
School of Civil Engineering
Qingdao University of Technology
Qingdao, China

ISSN 2366-2557 ISSN 2366-2565 (electronic)
Lecture Notes in Civil Engineering
ISBN 978-981-33-4323-8 ISBN 978-981-33-4324-5 (eBook)
https://doi.org/10.1007/978-981-33-4324-5

This Springer imprint is published by the registered company Springer Nature Singapore Pte Ltd.
The registered company address is: 152 Beach Road, #21-01/04 Gateway East, Singapore 189721, Singapore

Foreword

Dear colleagues and friends!

Firstly, I wanted to congratulate the representatives of the universities of India and China for holding a series of webinars on geotechnics and construction.

I participated in several of them as a Guest Speaker and as a member.

I would like to highlight the young age of lecturers from India and China, who gave very interesting, useful, informative lectures. In addition, there was interesting oral discussion and communication among participants after each lecture.

I would like to wish the young scientists the following for motivation and achievement of results:

To pay particular attention to the conduct of engineering and geological research, since the quality of these studies ensures the safety of the operation of buildings and structures in a real and distant future. It's kind of an investment in your building under construction.

Always with a positive or negative geotechnical situation, you yourself need to look at it at the construction site, since personal participation in the scientific approach and the analysis of the geotechnical situation are very important for a real assessment of what is happening

When starting a private business, I recommend that young people do not send income and earned money to their salaries, but send them, as far as possible, to the development of technology, the purchase of modern equipment, software, as well as advanced training for your employees.

With compliments,
Prof. A. Zh. Zhusupbekov
Director of the Geotechnical Institute
Immediate Past Vice President of ISSMGE for Asia

Preface

Indo-China Research Webinar Series in Civil and Environmental Engineering is jointly initiated by faculties from Shantou University, China, SVNIT, Surat and Indian Geotechnical Society Surat Chapter, Vardhaman College of Engineering, Hyderabad, and Qingdao University of Technology, China. The aim is to connect and motivate young scholars from India and China.

The first Indo-China Research Webinar Series in Civil and Environmental Engineering was held from 8 to 19 May, 2020. This workshop was collaboratively organized by Dr. Ankit Garg, Associate Professor, Shantou University, China, Prof. (Dr.) C. H. Solanki, Professor, SVNIT, Surat and Indian Geotechnical Society Surat Chapter, Gujarat, India, Dr. Chandra Bogireddy, Assistant Professor, Vardhaman College of Engineering, Hyderabad, India, and Dr. Junwei Liu, Vice Dean and Associate Professor, Qingdao University of Technology, China.

In the Indo-China research webinar, around 16 technical sessions were held with eight Speakers each from Universities in China and India. The total of participants in all the sessions combined were more than 5000. In total, ten Special Guests were invited that graced the audience with their motivational words promoting such cooperation in future. Guests include Prof. G. L. Sivakumar Babu, President, Indian Geotechnical Society and Professor IISc Bangalore, Prof. S. R. Gandhi, Director, S. V. NIT Surat, India, Prof. Gautam Biswas, Former Director, IIT Guwahati, Fellow of ASME, Prof. Da Hsuan Feng, Fellow American Physical Society, USA, Chairman of Advisory Board, Hainan University, Former Vice President for Research, University of Texas, USA, Prof. S. K. Das, Director, IIT Ropar, Prof. D. N. Singh, Institute Chair Professor, IIT Bombay and Editor in Chief, Environmental Geotechnics, Prof. Chandan Ghosh, IIT Jammu, NIDM, Ministry of Home affairs, Prof. Askar Zhussupbekov, Past Vice-President of ISSMGE for Asia and Director of the Geotechnical Institute. Prof. Neelamani, Senior Research Scientist, Kuwait Institute of Scientific Research, Dr. Jaykumar Shukla, Principal Engineer, Geo Dynamics, India, and Mr. Subba Reddy Nelluru (UK). Also, the webinar was supported by Dr. S. Sai Satyanarayana Reddy, Principal, VCE and Dr. K. Mallikharjuna Babu, Director and CEO, Vardhaman College of Engineering, Hyderabad.

We would like to thank Mengchu Huang, Karthikeyan Durairaj, and the whole Springer team for their full support and cooperation at various stages of the preparation and production of this proceedings book.

Shantou, China Ankit Garg
Surat, India C. H. Solanki
Hyderabad, India Chandra Bogireddy
Qingdao, China Junwei Liu

Contents

About the Editors

Dr. Ankit Garg is an Associate Professor in Department of Civil and Environmental Engineering at Shantou University. Prior to it, he obtained his Ph.D. from Hong Kong University of Science and Technology under Hong Kong Ph.D. Fellowship Scheme and B.Tech. from IIT Guwahati (2006–2010). He was former Assistant Professor at IIT, Guwahati (2015–2017), and also part time World Bank Consultant (2016–2017) for monitoring transport infrastructure projects in Assam, India. His research focus on the development and utilization of sustainable materials (biochar, fiber) includes vegetation for soil remediation. He has published more than 100 SCI publications, (including Canadian Geotechnical Journal, Geotechnical Testing Journal, ASTM, Geotechnique letters, and Journal of Hydrology) and also published three ESI papers among which two papers are in top 0.1%. He is awarded Telford Premium Prize from British Civil Engineers Association for his publication in Geotechnique letters. He is also currently Guest Editor for Special Issues related to Bio-engineering applications in soil and water remediation in "Journal of Hazardous Toxic and Radioactive Waste, ASCE", International Journal of Damage Mechanics (SCI Journal), International Journal of Geosynthetics and Ground Engineering, Springer (Web of Science) and Biomass Conversion and Bio-refinery (Springer, SCI) jointly with expertise from environmental science and soil science. Besides, he was awarded Talented Youth Scientist Program from MOST, China, Young Doctor Award from Ministry of Education Guangdong, and International Scientist Exchange Program. Further, he currently presides over National Natural Science Foundation of China Youth Project on "Vegetation for soil remediation". He has been invited to give lectures at Gifu University in Japan, Ruhr University in Germany, University of Leeds, UK, and National Ilan University in Taiwan.

Prof. C. H. Solanki is a Professor (Geotechnical Engineering) in Civil Engineering Department, S. V. National Institute of Technology, (NIT Surat), Gujarat, and faculty member since August 1991. Professor C. H. Solanki has guided 40 Postgraduate Dissertations and 16 Ph.D. Thesis and presently, Research Supervisor of 11 ongoing Ph.D. scholars. He has published more than 150 research papers in reputed international/national journals and conferences. He has received the "Shri. M. S. Jain Biannual Award" for Best Paper in IGC-2013 and the "Distinguished Faculty Award" from The Venus International Faculty Awards—2016, and also received two best paper awards in conferences. Professor C. H. Solanki has organized 16 national level events including STTP and seminars. He was the chairman for Indian Geotechnical Conference (IGC) 2019, Surat. He has been elected as Member of the Executive Committee IGS for the three terms (2015–2020). He worked as the Dean (Planning & Development) at SVNIT, Surat. He has published three books and edited proceedings. He has 29 years of teaching experience for undergraduate and postgraduate students.

Dr. Chandra Bogireddy is an Assistant Professor in Department of Civil Engineering, Geotechnical Engineering Division at Vardhaman College of Engineering, Hyderabad. He obtained his B.Tech. in Civil Engineering from Sri Kalahasteeswara Institute of Technology, Srikalahasti, affiliated to Jawaharlal Nehru Technological University, Anantapur, Andhra Pradesh, in the year 2010. Later, he worked as a Junior Engineer at Agri Gold Pvt. Ltd., Hyderabad, A.P., 2010. In 2012, he completed his M.Tech. in Geotechnical Engineering (Class Topper) from Sri Venkateswara University College of Engineering, S. V. University, Tirupati, Andhra Pradesh. During his master's dissertation, he worked on "Analysis of Engineering Behavior of Residual Tropical Soils-A Critical State Framework". Later, he worked as an Assistant Professor at Department of Civil Engineering, Sree Vidyanikethan Engineering College (Autonomous), Tirupati, A.P. from 2013–2014. He was awarded Ph.D. degree in 2019 from Department of Civil Engineering, SVNIT Surat, India. His Ph.D. research on "Rapid Assessment of Compressibility Behaviour of Clayey Soil from Experimental and Finite Element (FE) Simulations" also led to two best paper awards in National conferences. He delivered nine guest lectures in various national conferences/workshops. He was also chairman of technical session and evaluator for poster session at Indian Geotechnical Conference (GeoINDUS) 2019 organized by IGS Surat Chapter. He is also an Associate Member of ASCE, SM Institution of Civil Engineers, Institute of Engineers (India), and Fellow of FSIESRP Life Member. His areas of interest include basic Geomechanics, Numerical modelling in Geomechanics, Image analysis, and Site characterization.

Dr. Junwei Liu holds the position of Associate Dean in School of Civil Engineering at Qingdao University of Technology (QUT). He earned his Ph.D. in Civil Engineering from Zhejiang University and served as a post-doctoral scholar before joining QUT. His research interests include offshore foundation dynamics, disaster prevention and mitigation engineering, and soil-structure interaction mechanism. He also visited as a research scholar at Cambridge University, University College London and University of Surrey, for developing a strong collaboration and working in various intricate projects. Dr. Liu has published a textbook with 12 other articles in SCI indexed journals and 31 articles in EI indexed journals. He also authorized 17 invention patents and owns copyrights for 25 practical new-type patents and software. Dr. Liu presided two projects for Natural Science Foundation of China and 11 other projects in provincial and ministerial levels. The project includes Metro construction for the city of Qingdao and Hangzhou and airport construction at Jiaodong. Dr. Liu has been awarded 8 academic level awards and seven other major awards, securing first prizes at Qingdao S&T Progress, China Business Federation Science and Technology Progress, and Heilongjiang Provincial Science & Technology Progress. As a member, he serves as an academic leader for "Shandong Provincial College Youth Innovation Team" and in other institutions including "Geotechnical Earthquake Engineering Committee of China", "China Association for Engineering Construction Standardization", "The Institution of Civil Engineering (ICE)", and "Shandong Province Civil Engineering Society".

Effect of Earthquake on Underground Metro Tunnels—A Parametric Study

Manendra Singh, N. K. Samadhiya, and Jitendra Singh Yadav

Abstract Metro underground tunnels are life line in many major cities across the world. Therefore vulnerability of these structures against the earthquakes is very critical. In this chapter, Seismic analysis of Delhi metro underground tunnels have been performed by finite element method. Chamoli earthquake (1999) of lower Himalaya, India has been taken for the analysis. Parametric study was also conducted by considering the parameters like volume loss during construction, depth of overburden, interfacial characteristics between soil and the tunnel liner, thickness of RC liners, peak ground acceleration, and damping. Values of the vertical stress concentration were found to increase with increase in volume loss during the construction. This observation is true for both elastic and elasto-plastic behavior of soil. Maximum dynamic axial force in RC liners has been found to increase with the ratio of depth of overburden to the tunnel size, whereas both shear force and bending moment in lining of tunnel have been found to reduce as this ratio attains value of 2 and 2.5 respectively. Dynamic forces in RC liners tend to reduce as the value of interaction factor at the interface between the tunnel and RC liners reduces from 1 to 0.1. Increasing the thickness of RC liners causes the axial force in RC liners to reduce but bending moment has been found to increase. These points need to be borne in mind while designing the support systems for the tunnels.

Keywords Seismic behavior · Metro underground tunnels · Earthquake · Volume loss

M. Singh (✉) · J. S. Yadav
Department of Civil Engineering, National Institute of Technology Hamirpur, Hamirpur, HP, India
e-mail: manendra@nith.ac.in

N. K. Samadhiya
Department of Civil Engineering, Indian Institute of Technology Roorkee, Roorkee, India

A. Garg et al. (eds.), *Proceedings of the 1st Indo-China Research Series in Geotechnical and Geoenvironmental Engineering*, Lecture Notes in Civil Engineering 123,
https://doi.org/10.1007/978-981-33-4324-5_1

1 Introduction

With ever increasing population and hence traffic on roads in many cities, India is going ahead with major infra-structure development projects including the Mass Rapid Transit System (MRTS) for metropolitan cities like Delhi, Kolkata, Bangalore, Hyderabad, Lucknow, Jaipur, etc. Metro underground tunnel network has already been constructed on a wide scale in Delhi and it is also advancing to the remaining parts of city. Metro underground tunnels constructed in Delhi are mostly at shallow depth and these have to be safe against any earthquake. In the event of an earthquake, any damage to these shallow tunnels can have a significant economic impact apart from the aspect of the Non Performance Time (NPT) of that network and hence lot of inconvenience to the general public. Considering these issues, this study is focused on exploring the earthquake effect on seismic response of metro underground tunnels. Singh et al. [1, 2] performed seismic analysis of Delhi metro underground tunnels for their linear elastic response to 1999 Chamoli and 1991 Uttarkashi earthquakes of lower Himalaya. The response of tunnels was evaluated in terms of time histories of displacement, acceleration and forces in RC liners. Singh et al. [3] also studied the elasto-plastic behavior of Delhi metro tunnels to both Chamoli and Uttarkashi earthquakes and predicted time histories of displacement and acceleration in soil-tunnel system and axial force, shear force and bending moments in RC liners. In each of the above studies, the numerically predicted results were compared with the closed form solutions of Wang [4], Penzien et al. [5, 6]. In this chapter, Seismic analysis of Delhi metro underground tunnels has been performed by finite element method using Plaxis software in order to understand the influence of volumes loss that might occur during the construction of tunnels, apart from the influence of significant parameters like depth of overburden, nature of interface between soil and tunnel liners, thickness of RC liners, peak ground acceleration, and damping of soil.

2 Earlier Work

Various authors, namely Dowding and Rozen [7], Owen and Scholl [8], Sharma and Judd [9], Rowe [10], Power et al. [11], Hashash et al. [12] and Lanzano et al. [13] have studied the causes of failure, damage pattern, and behavior of underground structures during the earthquakes. It was concluded from these studies that the damage reduced with increasing depth of overburden. Structures which were buried in soft soils experienced more damage than those of structures in rock. Wang [4], Penzien et al. [5, 6], Hashash et al. [12], and Park et al. [14] have presented closed form solutions for seismic analysis of underground tunnels. There are various limitations of these closed formed solutions. Therefore numerical analysis is very important for predicting the realistic behavior of underground tunnels during the earthquake. Pakbaza and Yareevand [15] showed that there was good agreement between maximum bending moment and maximum axial thrust between closed form and numerical solutions. Amorosi and Boldini [16] investigated the dynamic behavior of circular tunnels in clayey soils (soft and stiff clay) in the transverse direction. The results of visco-elastic analysis and Wang's [4] solutions were found to be comparable, especially for the soft clay. Forces in tunnel lining were modified due to plasticity-based analyses. Permanent increments of both hoop force and the bending moment also occurred in plasticity-based analyses. Shahrour et al. [17] studied the effect of soil plasticity and soil dilatancy on the seismic response of underground tunnels. Beshrat et al. [18] have shown that presence of tunnel modifies the behavior of the medium and also behavior at the ground surface. The amplification of acceleration was more in weaker soil than the harder one. It was therefore proposed that in urban areas, in adjacent underground structures like tunnels, the peak ground acceleration (PGA) should be chosen from dynamic analysis in presence of tunnel instead of using the codes. Beshrat et al. [19] concluded that internal compressive pressure increases with increase in peak ground acceleration of earthquake. Sahoo and Kumar [20] observed that residual and maximum forces in the liners of tunnel were highly influenced by peak ground acceleration. Cilingir and Gopal Madabhushi [21] concluded that safety of underground tunnels against the earthquake damage increases with increasing depth of overburden. Liu and Song [22] observed that maximum distortion in tunnel lining depends upon the relative stiffness between tunnel and the surrounding soil. Moreover, distortion in lining of tunnel is highly dependent on peak ground acceleration (PGA) and the overburden depth of tunnel.

3 Problem Definition

Metro underground trains in the capital city of Delhi provide a major mass rapid transit system. In this chapter, a typical section of underground tunnel of Delhi Metro has been considered for the analysis. Properties of soil and the tunnel have been adopted from Yadav [23] and Soni [24]. Table 1 shows the details of geometric parameters of tunnel. Attempt has been made to analyse the tunnel section in detail for Chamoli earthquake (1999) of the lower Himalaya. Alluvium deposits of soil was

Table 1 Properties of tunnel

Properties	Values
Tunnel's diameter (D)	6.26 m
Depth of overburden (H)	16.87 m
Support system of tunnel	Segmental reinforced concrete (RC) liners
Thickness RC liners	0.28 m
Elastic modulus of lining of tunnel (E_c)	3.16×10^7 kN/m^2
Poisson's ratio	0.15
Damping ratio	2%

Table 2 Detail of soil properties

Properties	Values
Unit weight of soil (γ_{bulk})	18 kN/m^3
Saturated unit weight of soil (γ_{sat})	20 kN/m^3
Cohesion ©	0
Angle of friction (φ)	35°
Dilatational angle (ψ)	5°
Poisson's ratio	0.25
Damping ratio	15%

Table 3 Variation of elastic modulus of Delhi silt with depth [20, 21]

Depth (m)	Thickness (m)	Elastic modulus (kPa)
0–10	10	7500
10–20	10	15,000
20–35	15	30,000
35–50	15	40,000
50–60	10	50,000

found to be exist under this tunnel section. Detail of the soil properties surrounding the tunnel have been tabulated in Table 2. These properties have been kept constant with depth of soil whereas elastic modulus of soil varies with depth. This variation of elastic modulus with depth is presented in Table 3. Depth of water table was well below the soil tunnel system. Schematic diagram of soil-tunnel system is shown in Fig. 1.

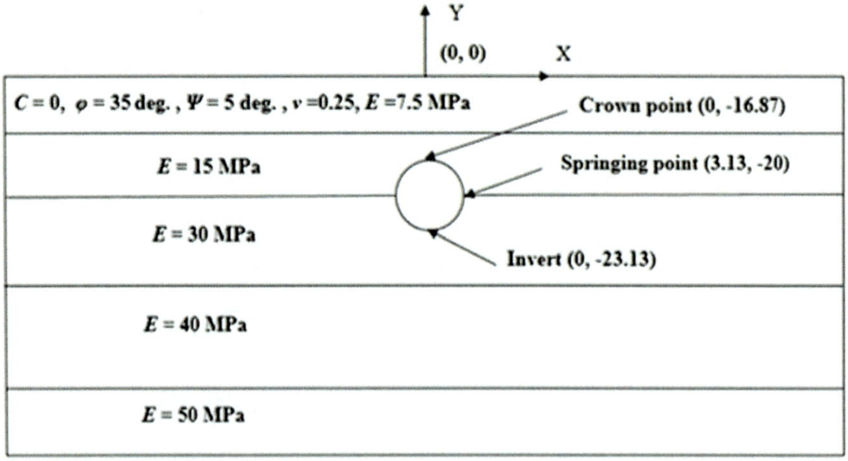

Fig. 1 Schematic diagram of soil-tunnel system

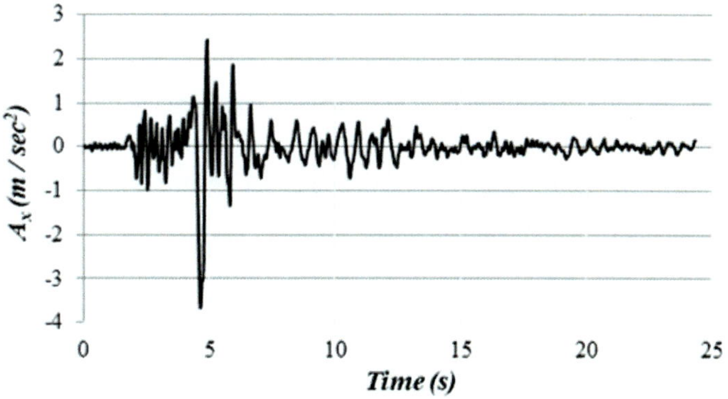

Fig. 2 Time history of horizontal acceleration (Ax)—(Chamoli earthquake, 1999, India)

4 Numerical Modeling

2D plane-strain finite element analysis of DMRC metro underground tunnel has been performed using PLAXIS 2D software. Horizontal domain of soil was taken as 140 m × 60 m in size. Soil medium was simulated using Mohr coulomb model whereas RC liners of tunnel was simulated by plate elements. No-slip condition has been assumed initially but subsequently, parametric study has been carried out with variation in values of R_{inter} from 1.0 to 0.01. Where R_{inter} is an interface factor. Friction angle and cohesion of soil and interface are dependent on this interface factor and calculated by Eqs. 1 and 2, respectively [25].

$$\tan \phi_{interface} = R_{inter} \tan \phi_{soil} \tag{1}$$

$$C_{interface} = R_{inter} C_{soil} \tag{2}$$

Rayleigh damping is simulated in Plaxis software. The damping of soil and RC liners of tunnels were 15% and 2%, respectively. For static response, elementary boundary was used whereas for dynamic viscous absorbent boundary as given by Lysmer and Kuhlmeyer [26] was used. Chamoli earthquake (1999) was used for the seismic analysis. The acceleration versus time history of horizontal component of this earthquake has been shown in Fig. 2. Time integration scheme as given by Newmark [27] was employed in this study.

5 Results and Discussions

In this section effects of volume loss are discussed in terms of vertical stress concentration and forces in tunnel liner for static as well as dynamic analysis.

5.1 Static Analysis

Static analysis was performed as an initial step. Contraction of 1–4% was taken for simulating the volume loss during the construction of tunnel. Results of this analysis are presented and discussed below in the form of vertical stress concentration, vertical displacement and the forces in RC liners.

5.1.1 Vertical Stress Concentration

Vertical stress concentration is the ratio of vertical stress at any point after tunnel construction $(\sigma_s)_{yy}$ to the initial stress $(\sigma_i)_{yy}$ at that point. Figure 3 shows the variation of vertical stress concentration along the horizontal axis of tunnel for different values of volume contraction, for linear elastic behavior of soil, where a is the radius of the tunnel and r is the radial distance along the horizontal axis of tunnel from the tunnel centre. It can be noticed that stress concentration is maximum at springing points ($r/a = 1$) and which decreases up to a concentration factor of almost unity at $r/a = 6$, then thereafter remains constant. It can be also observed from Fig. 3, that stress concentration at $r/a = 1$, increases with the increase in volume loss, especially near the springing point. These stress concentration factor just near the springing point was found to be 1.78, 1.58, 1.42 and 1.19 for volume contraction of 4%, 3%, 2% and 1% respectively. So the effect of volume contraction should be accounted for carefully. Similarly Fig. 4 shows the variation of stress concentration along the horizontal axis of the tunnel for elasto-plastic behavior of soil. It can be seen that as the soil mass immediately surrounding the tunnel enters into plastic state, stress concentration factor drops down to a value less than unity at the tunnel periphery and then increases to a maximum value at a certain distance away from the periphery, called as the elasto-plastic radius, which in this case is 1.5 times the tunnel radius. Beyond this point, the stress concentration factor consistently drops down to unity at

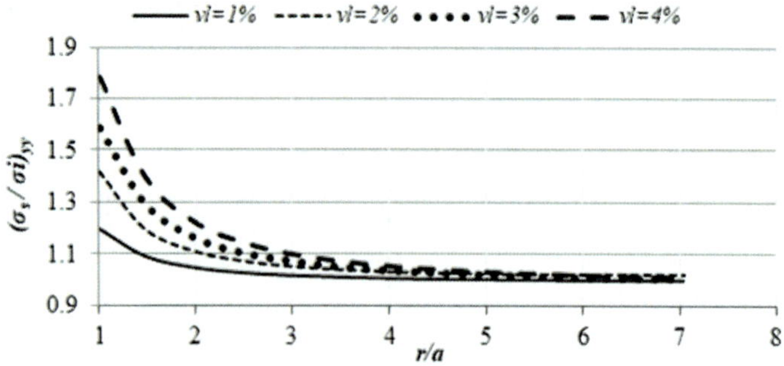

Fig. 3 Elastic stress concentration along horizontal axis of tunnel

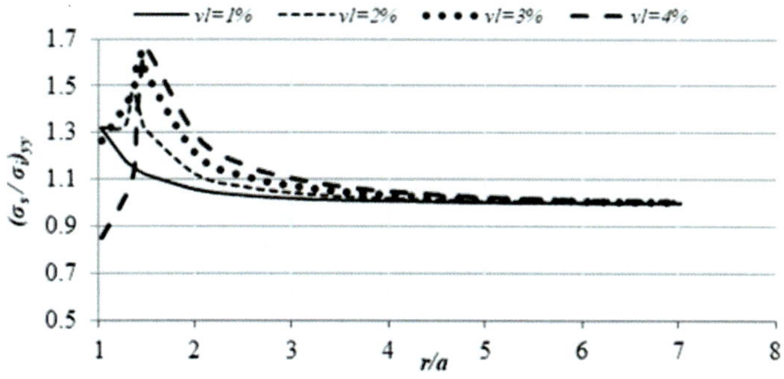

Fig. 4 Elasto-plastic stress concentration along horizontal axis of tunnel

Table 4 Vertical displacement (mm) at the ground surface due to construction of tunnel

V_l (%)	U_y (linearelastic analysis) (mm)	U_y (elasto-plastic analysis) (mm)
1	1.396	3.867
2	6.91	14
3	15	27
4	24	40
V_l (%)	U_y (linearelastic analysis) (mm)	U_y (elasto-plastic analysis) (mm)

a distance far away from the tunnel periphery. It can also be observed that for volume loss of 1%, stress concentration profile follows the conventional trend for an elastic analysis. Therefore, volume loss for non-linear analysis should be taken between 2 and 4%.

5.1.2 Vertical Displacement at Ground Surface

Vertical displacement at ground surface, known as the ground subsidence, occurs during the construction of tunnel. Values of vertical displacement obtained at the ground surface are presented in Table 4 which shows that subsidence at ground surface increases with the increase in volume contraction, both for linear elastic and elasto-plastic behavior of soil. Moreover, displacement at ground surface increases due to non-linearity of soil mass.

5.1.3 Forces in RC Liners

Table 5 shows the variation of maximum axial force (T_{max}), shear force (V_{max}) and bending moment (M_{max}) in RC liners of tunnel with increase in volume loss. It is obvious that axial force decreases significantly with increase in volume loss (V_l) for

Table 5 Maximum forces in RC liners in static condition

Volume loss, Vl (%)	Linear elastic behavior of soil			Elasto-plastic behavior of soil		
	$T_{\max}$ (kN/m)	$V_{\max}$ (kN/m)	$M_{\max}$ (kNm/m)	$T_{\max}$ (kN/m)	$V_{\max}$ (kN/m)	$M_{\max}$ (kNm/m)
1	872.51	23.00	12.74	798.83	174.33	271.85
2	638.22	42.61	23.79	635.43	173.89	267.15
3	438.64	62.22	39.40	527.50	171.99	259.90
4	239.06	81.83	55.01	459.97	164.18	274.09

both linear elastic and elasto-plastic behavior of soil. It can be also observed that both shear force and bending moment increase with increase in volume loss in case the soil behaves in an elastic manner. However, in case of elasto-plastic soil behavior, both shear force and bending moment were found to be less affected by the increase in volume loss.

5.2 *Dynamic Analysis*

The analysis was performed for the horizontal component of 1999 Chamoli earthquake of the lower Himalaya. Earthquake was applied at the base of model.

5.2.1 Dynamic Vertical Stress Concentration

Figure 5 shows the dynamic stress concentration variation along the horizontal axis of the tunnel for different values of volume contraction during the earthquake, for the elastic behavior of soil. Dynamic stress concentration factor is the ratio of σ_d, the maximum vertical stress during earthquake to σ_s, the static vertical stress after construction of tunnel. Figure 6 shows similar variation of stress concentration factor during earthquake for the elasto-plastic behavior of soil. It can be seen from the both plots that dynamic vertical stress concentration factor is independent of the volume loss.

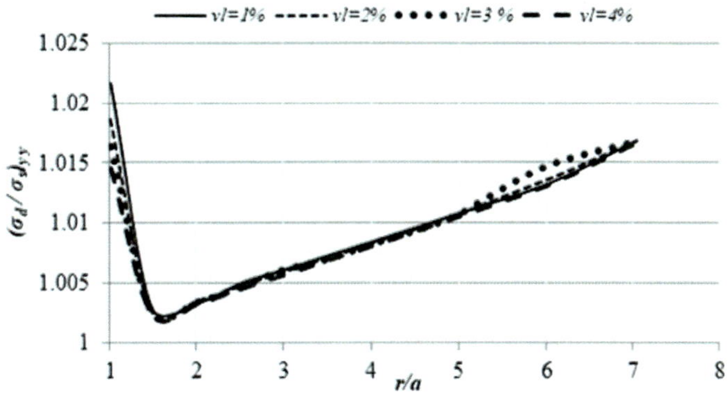

Fig. 5 Dynamic stress concentration along horizontal axis of tunnel during 1999 Chamoli earthquake (elastic analysis)

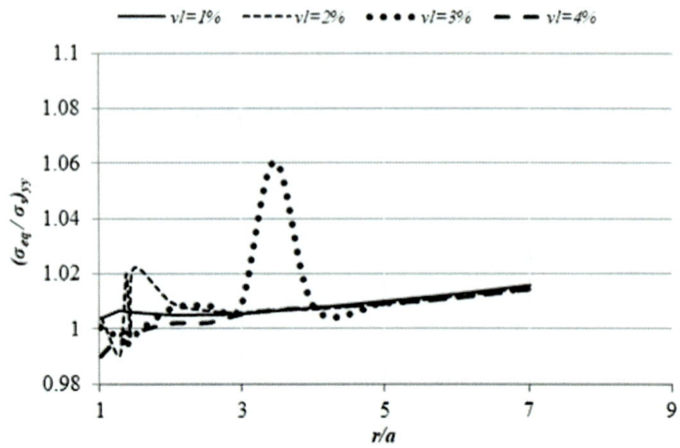

Fig. 6 Dynamic stress concentration along horizontal axis of tunnel during 1999 Chamoli earthquake (elasto-plastic analysis)

5.2.2 Forces in RC Liners

Table 6 shows the variation of incremental forces in RC liners during the earthquake with the volume loss. It can be seen that axial force increases significantly whereas the bending moment decreases with increase in volume loss (V_l) and is independent of state of stress soil i.e. elastic or plastic. Shear force in RC liners remains almost constant irrespective of whether soil is in elastic or plastic state.

5.2.3 Parametric Study

Attempt has now been made to conduct some parametric study, especially the influence of: (i) Overburden depth, (ii) Interface between soil and tunnel liner, (iii) Thickness of RC liners, (iv) Peak ground acceleration, (v) Boundary conditions, and (vi) Damping. The study has been carried out for the horizontal component of Chamoli earthquakes with 15% damping ratio for the soil. Contraction of 3% was taken for simulating the volume loss during the tunnel construction. All other properties remain the same.

Influence of Overburden Depth

For studying the effect of overburden depth of tunnel, homogeneous soil mass is adopted, with elastic modulus 29.58 MPa and 15% damping of soil. Other properties are kept same as above. The variation of axial force, shear force and bending moment in the liners is plotted in Fig. 7a–c with non-dimensional depth of overburden. It can be noticed that maximum axial force increases with H/D ratio whereas both shear

Table 6 Maximum forces in RC liners during 1999 Chamoli earthquake

V_l (%)	Considering linear elastic behavior of soil			Considering elasto-plastic behavior of soil		
	δT_{max} (kN/m)	δV_{max} (kN/m)	δM_{max} (kNm/m)	δT_{max} (kN/m)	δV_{max} (kN/m)	δM_{max} (kNm/m)
1	287.49	87.28	129.06	221.17	26.38	38.01
2	521.78	87.04	121.65	384.57	22.95	37.93
3	721.36	87.04	109.7	492.5	21.4	33.76
4	920.94	87.04	98.93	560.03	26.26	5.88

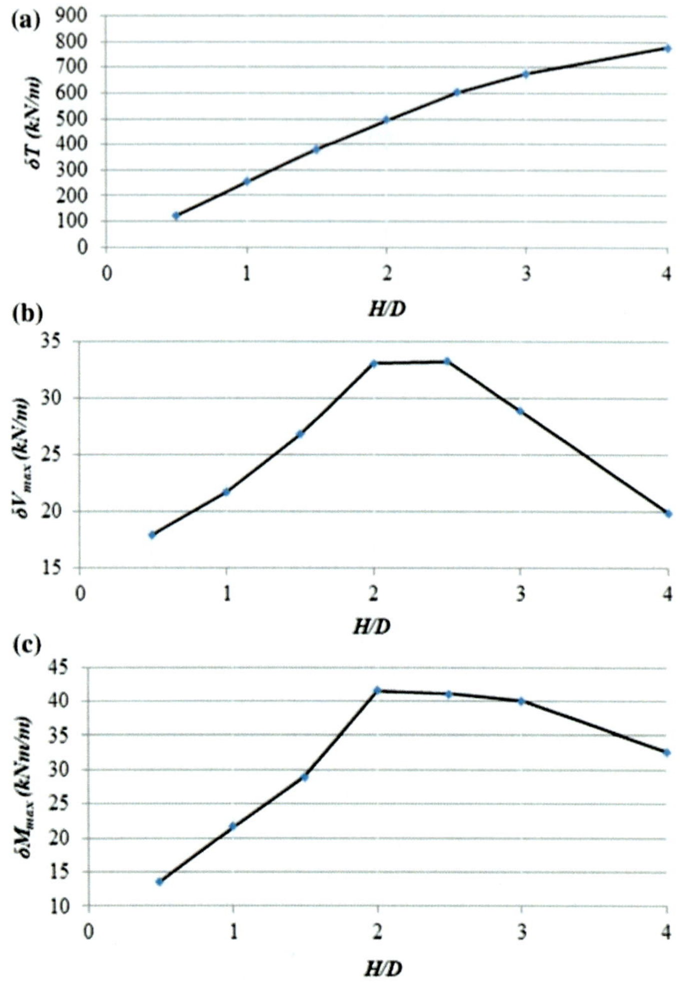

Fig. 7 Effect of overburden depth on **a** maximum axial force, **b** maximum shear force, **c** maximum bending moment due to 1999 Chamoli earthquake

force and bending moment initially increase up to H/D ratio of 2 and then show a decreasing trend.

Effect of Interface Between Soil and Tunnel Liners

Soil-tunnel system has been carried out with 15% damping ratio of soil and with overburden depth of 16.87 m. For no slip condition, Plaxis defines a parameter, R_{inter} as 1.0. As R_{inter} reduced, relative slip between soil and tunnel liner is allowed. Different values of R_{inter} = 1, 0.5, 0.3, 0.2, 0.1, and 0.01 are considered in the

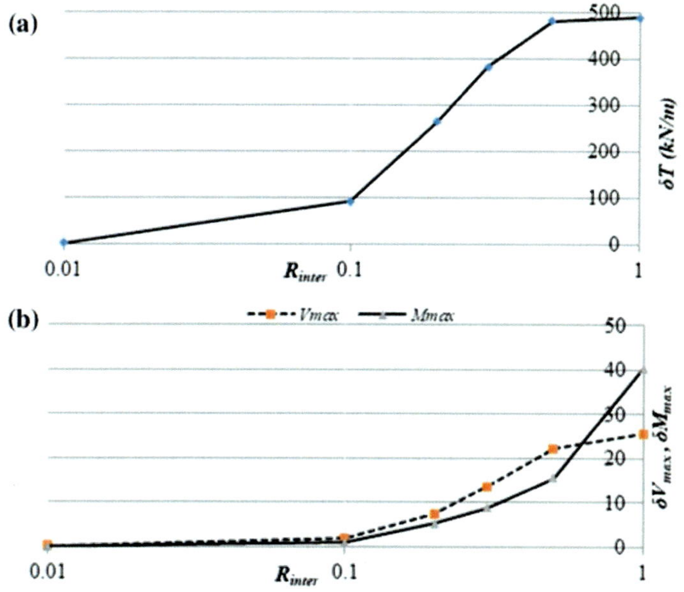

Fig. 8 Effect of interface characteristics between soil and tunnel liners on **a** maximum axial force, **b** maximum shear force and bending (1999 Chamoli earthquake)

analysis. Plots in Fig. 8a, b show the variation of dynamic forces and bending moment developed in RC liners with the interface factor, R_{inter}. Expectedly, the dynamic forces and bending moment attain maximum their values at $R_{inter} = 1$ i.e. when no relative slip is permitted between the liners and the surrounding soil mass. As R_{inter} reduces from no slip to partial slip condition, to a value of 0.1, there is a drastic resduction in liner forces and bending moment. However, with further reduction of R_{inter} to a value of 0.01, i.e. when almost the full slip condition is reached, the reduction in forces and moments is very gradual.

Effect of Thickness of Liners

Analysis was further carried out for different values of liner thickness, *t* varying from 0.05 m to 2.0 m. This was done with 15% damping ratio of soil and with overburden depth of 16.87 m. The outcome of analysis for Chamoli earthquake is clear in Fig. 9a–c. A sudden reduction in maximum axial force in liners can be observed up to a liner thickness of 1.0 m beyond which axial force remains practically constant (Fig. 9a). The maximum shear force increases initially up to a liner thickness of 0.60 m beyond which, it reduces a bit but almost remains constant (Fig. 9b). The maximum bending moment in the liners increases consistently with almost a linear trend (Fig. 9c). Therefore increasing the thickness of liners does not give any advantage as far as the maximum bending moment is concerned.

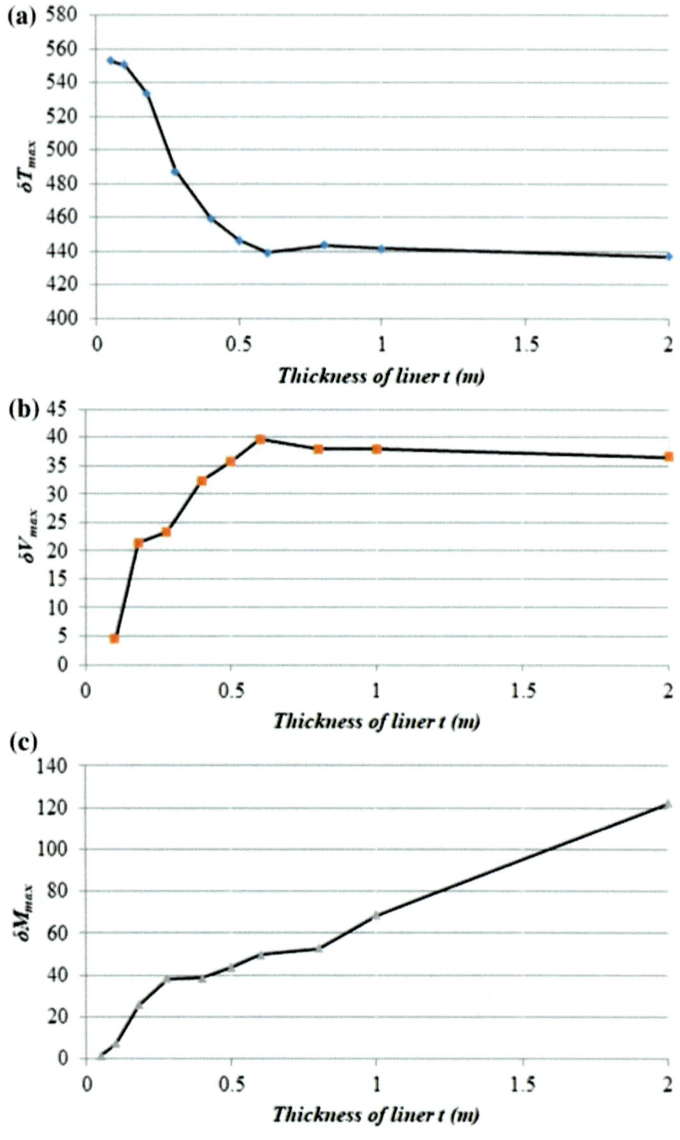

Fig. 9 Effect of thickness of tunnel liners on **a** maximum axial force, **b** maximum shear force, **c** maximum bending moment (1999 Chamoli earthquake)

Influence of Peak Ground Acceleration

Figure 10a, b show the variation of dynamic forces and moments with the peak ground acceleration. In general, it has been found that residual dynamic axial force in RC liners show consistently an increasing trend with increase in peak ground acceleration

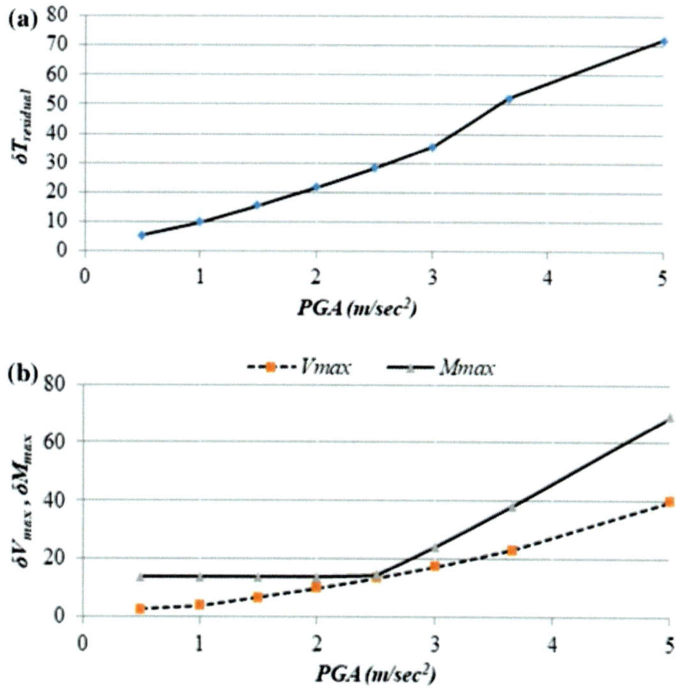

Fig. 10 Effect of peak ground acceleration on **a** residual axial force, **b** maximum shear force and bending moment in RC liners (1999 Chamoli earthquake)

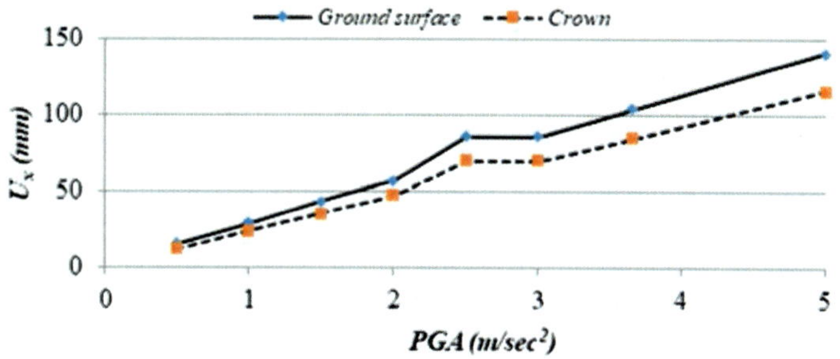

Fig. 11 Effect of PGA on horizontal displacement (1999 Chamoli earthquake)

(Fig. 10a). The maximum shear force in liners remains constant up to a value of PGA of 2.0 which then continuously increases till PGA attains a value of 5.0 (Fig. 10b). The maximum bending moment in liners, however, shows a consistent increasing trend (Fig. 10b). Effect of PGA on horizontal displacement at ground surface and tunnel crown has been plotted in Fig. 11. Although horizontal displacement shows an increasing trend throughout, it remains constant over a range of PGA between 2.5 and 3.0.

Table 7 Effect of damping on forces in RC liners during 1999 Chamoli earthquake

Maximum forces in RC liners	Static case	During 1999 Chamoli earthquake (absorbent boundaries)				15% damping with nonabsorbent boundaries
		0% damping	5% damping	10% damping	15% damping	
T (kN/m)	527.5	1020	1020	1020	1020	1020
V (kN/m)	171.9	252.16	216.89	202.62	193.39	195.80
M (kNm/m)	259.9	369.5	330.98	315.36	293.66	297.32

Effect of Boundary Conditions

The influence on forces mobilized in RC liners is shown in Table 7. In case of non-absorbent boundaries, shear force and bending moment in RC liners were slightly higher due to reflection of waves and hence absorbent boundaries should be considered in any seismic analysis.

Influence of Damping

Effect of damping was found out in terms of acceleration and displacement. Figure 12a shows the maximum horizontal displacement–time history observed at the ground surface due to application of horizontal component of Chamoli earthquake for different values of damping ratio. The maximum horizontal displacement is 216.91 mm for 0% damping ratio, which then consistently reduces to 132.53, 105.29 and 103.19 mm with increase in damping ratio from 5 to 15%. Figure 12b, which shows similar time history of horizontal acceleration at ground surface, and gives values of maximum horizontal acceleration of 6.23, 3.74, 2.38, and 1.45 m/s^2 for increasing values of damping ratio. Therefore maxima values of both horizontal displacement and horizontal acceleration reduce with increase in soil damping. Influence of soil damping on forces mobilized in RC liners is presented in Table 7. A close look at Table 7 suggests that axial force in RC liners is not at all affected with increase in damping, whereas both shear force and bending moment reduce significantly with increase in soil damping.

6 Conclusions

(i) Static analysis gives maximum vertical stress concentration during the construction of tunnel at the springing points along the tunnel periphery (i.e. $r/a = 1.0$), when state of stress in soil mass elastic whereas maximum vertical stress concentration occurs during tunnel construction at the elasto-plastic radius (i.e. $r = r_p$) when soil response is elasto-plastic.

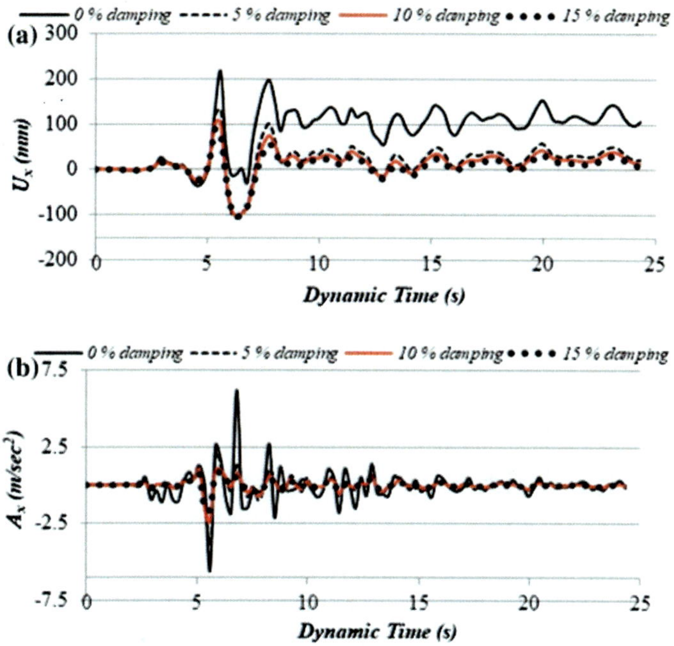

Fig. 12 **a** Horizontal displacement–time history **b** horizontal acceleration-time history at ground surface for 1999 Chamoli earthquake as a function of damping ratio

(ii) The vertical stress concentration during tunnel construction is significantly affected by the tunnel volume loss. Similarly, increase in volume loss causes consistent reduction of axial force in RC liners, whereas shear force and bending moment are marginally affected. This is true for both elastic and in-elastic soil response.

(iii) It has been found that dynamic axial force in RC liners increases significantly (by more than 3.2 times in elastic case and by more than 2.5 times in elasto-plastic situation) with increase in volume loss (V_l). Both dynamic shear force and dynamic bending moment in RC liners have been found to increase only marginally with increase in volume loss. This is true for both elastic and in-elastic situations

(iv) Maximum dynamic axial force in RC liners has been found to increase with increase in depth of overburden (H/D ratio), whereas both dynamic shear force and bending moment in RC liners increase initially up to the H/D ratio of 2.0 and then continue to reduce at higher values of H/D ratio.

(v) Dynamic forces in RC liners including axial force, shear force and bending moment continue to increase significantly but only for values of R_{inter} varying between 0.1 and 1.0.

(vi) Thickness of RC liners has a significant influence on dynamic axial force and shear force in RC liners up to a liner thickness of about 1.0 m, however, it is only the dynamic bending moment which is affected, in fact it increases linearly with increase in thickness. So increasing the thickness of RC liners is not a good choice from design point of view.

(vii) Peak ground acceleration exerts undoubtedly a significant influence on dynamic forces in RC liners which have been found to increase consistently.

(viii) Damping in soil mass significantly influences, not the axial force but both shear force and the bending moment in RC liners.

References

1. Abdel-Motaal MA, El-Nahhas FM, Khiry AT (2014) Mutual seismic interaction between tunnels and the surrounding granular soil. HBRC J 1(3): 265–278
2. Singh M, Viladkar MN, Samadhiya NK (2017) Seismic analysis of Delhi metro underground tunnels. Indian Geotech J 47(1):67–83
3. Singh M, Viladkar MN, Samadhiya NK (2017) Elasto-plastic analysis of Delhi metro underground tunnels under seismic condition. In: 16th world conference on earthquake engineering (16 WCEE), Santiago, Chile, 9–13 Jan 2017
4. Wang JN (1993) Seismic design of tunnels: a state-of-the-art approach. Monograph, monograph 7. Parsons, Brinckerhoff, Quade and Douglas Inc., New York
5. Penzien J, Ching LWu (1998) Stresses in linings of bored tunnels. Earthq Eng Struct Dyn 27(3):283–300
6. Penzien J (2000) Seismically induced racking of tunnel linings. Earthq Eng Struct Dyn 29(5):683–691
7. Dowding CH, Rozan A (1978) Damage to rock tunnels from earthquake shaking. ASCE J Geotech Eng Div 104(2):175–191
8. Owen GN, Scholl RE (1981) Earthquake engineering of large underground structures
9. Sharma S, Judd WR (1991) Underground opening damage from earthquakes. Eng Geol 30(3–4):263–276
10. Rowe R (1992) Tunnel engineering in earthquake area. Tunn Tunn 12:41–44
11. Power MS, Rosidi D, Kaneshiro J (1996) Strawman: screening, evaluation, and retrofit design of tunnels. Report draft, vol III. National Center for Earthquake Engineering Research, Buffalo, New York
12. Hashash YMA, Hook JJ, Schmidt B, John I, Yao C (2001) Seismic design and analysis of underground structures. Tunn Undergr Space Technol 16(4):247–293
13. Lanzano G, Bilotta E, Russo G, Silvestri F (2015) Experimental and numerical study on circular tunnels under seismic loading. Eur J Environ Civ Eng 19(5):539–563
14. Park K-H, Tantayopin K, Tontavanich B, Owatsiriwong A (2009) Analytical solution for seismic-induced ovaling of circular tunnel lining under no-slip interface conditions: a revisit. Tunn Undergr Space Technol 24(2):231–235
15. Pakbaz MC, Yareevand A (2005) 2-D analysis of circular tunnel against earthquake loading. Tunn Undergr Space Technol 20(5):411–417
16. Amorosi A, Boldini D (2009) Numerical modelling of the transverse dynamic behaviour of circular tunnels in clayey soils. Soil Dyn Earthq Eng 29(6):1059–1072
17. Shahrour I, Khoshnoudian F, Sadek M, Mroueh H (2010) Elastoplastic analysis of the seismic response of tunnels in soft soils. Tunn Undergr Space Technol 25(4):478–482
18. Besharat V, Davoodi M, Jafari MK (2012) Effect of underground structures on free-field ground motion during earthquakes. In: 15th world conference on earthquake engineering
19. Beshrat V, Davoodi M, Kazem M (2012) Effects of underground structures on free-field ground motion during earthquakes. In: The 15th world conference on earthquake engineering, Lisbon, Portugal, Sept 2012, pp 24–28

20. Sahoo JP, Kumar J (2014) Stability of a circular tunnel in presence of pseudostatic seismic body forces. Tunn Undergr Space Technol 42:264–276
21. Cilingir U, Gopal Madabhushi SP (2011) A model study on the effects of input motion on the seismic behaviour of tunnels. Soil Dyn Earthq Eng 31(3):452–462
22. Liu H, Song E (2005) Seismic response of large underground structures in liquefiable soils subjected to horizontal and vertical earthquake excitations. Comput Geotech 32(4):223–244
23. Yadav HR (2005) Geotechnical evaluation and analysis of Delhi metro tunnels. PhD dissertation
24. Sony S (2015) Static and dynamic response of Delhi metro tunnels. PhD dissertation. PhD thesis, IIT Delhi
25. Plaxis 2D version 8.0. Finite element codes for geotechnical engineering. https://www.plaxis.nl/
26. Lysmer J, Kuhlemeyer RL (1969) Finite dynamic model for infinite media. J Eng Mech Div 95(4):859–878
27. Newmark NM (1968) Problems in wave propagation in soil and rock. In: Proceedings of the international symposium on wave propagation and dynamic properties of earth materials, Univ. of New Mexico Press, Albuquerque, 23–25 Aug 1968, NM, pp 7–26

An Unfrozen Water Retention Curve for Capturing Soil Density and Specific Surface Effects

Jiwen Zhang and Qingyi Mu

Abstract Unfrozen water retention curve (UWRC) defines the relationship between temperature and unfrozen water content in frozen soils. Although many models have been proposed for the UWRC, these existing models cannot predict UWRC well over a wide temperatures range. In this study, a new UWRC model is proposed with explicit considerations of both capillarity and adsorption. In this model, capillarity is considered dominating when the freezing of soil pore water at higher temperatures (above −2 °C), whereas the effects of adsorption pronounce at temperatures below −2 °C. Moreover, effects of void ratio on the freezing of capillary water are incorporated. The proposed model was applied to predict UWRCs of silt and clay at different initial void ratios over a wide temperature range (from −50 to 0 °C). Predicted results by this new model are compared with predictions by three well-known existing models. The new model can capture the density effects on UWRC. Moreover, the new model can predict better UWRC over a wide temperature range since it explicitly considers both effects of capillarity and adsorption.

Keywords Unfrozen water retention curve · Frozen soil · Wide temperature range

1 Introduction

For solute-free soil, the capillarity and adsorption are the two distinct mechanisms controlling the entire freezing of pore water [1]. Unfrozen water retention curve (UWRC), defining the relationship between temperature and unfrozen water content,

J. Zhang · Q. Mu (✉)
Department of Civil Engineering, Xi'an Jiaotong University, 710049 Xi'an, Shanxi, China
e-mail: qingyimu@mail.xjtu.edu.cn

J. Zhang
JIKAN Research Institute of Engineering Investigations and Design, Co., Ltd., 710043 Xi'an, Shanxi, China

Shaanxi Key Laboratory for the Property and Treatment of Special Soil and Rock, 710043 Xi'an, Shanxi, China

A. Garg et al. (eds.), *Proceedings of the 1st Indo-China Research Series in Geotechnical and Geoenvironmental Engineering*, Lecture Notes in Civil Engineering 123,
https://doi.org/10.1007/978-981-33-4324-5_2

is an important soil property for calculating frost heave and thawing settlement [2, 3]. Many semi-empirical models were proposed for UWRC based on experimental data. Anderson and Tice [5] found that UWRC can be approximated as a power equation as follows:

$$\theta_w = \frac{p_d(1-\theta_s)}{100p_w}\alpha(-T)^{\beta} \quad (1)$$

where θ_w is the unfrozen volumetric water content; T is the soil temperature (°C); α and β are soil parameters which are related to specific surface area or liquid limit; p_d and p_w are the soil dry density and water density respectively and θ_s is the saturated volumetric water content. Equation (1) is formulated on the assumption that the adsorptive rather than capillary force governs the freezing of pore water. On the other hand, the exponential equation was also widely used for UWRC [6]:

$$\theta_w = \theta_{res} + (\theta_s - \theta_{res})\exp(\alpha_1 T) \quad (2)$$

where θ_{res} represents the residual volumetric water content; α_1 is a soil parameter defining the freezing rate of pore water. The freezing of residual water is ignored in this model.

Alternatively, some UWRC models have been developed on the basis of soil water retention models and the Clausius-Clapeyron equation. The water retention curve proposed by van-Genuchten [7] was used by Nishimura et al. [8]:

$$\theta_w = \left[1 + \left(\frac{\lambda p_w \ln((T+273.15)/273.15)}{\alpha_0}\right)^{n_0}\right]^{-m_0} \quad (3)$$

where α_0, n_0 and m_0 are soil parameters and λ is the latent heat of fusion of water (333.7 kJ kg^{-1} at 0 °C). In addition, the Fredlund and Xing equation [9] were also proposed to formulate the UWRC. However, these models may not give a well calculation for the UWRC over a wide range of temperatures as the lack of considerations of both capillarity and adsorption. Moreover, effects of initial void ratio, which significantly influences the capillarity [9, 10], on UWRC was not explicitly considered in these models.

A new UWRC model is proposed, which separates the freezing of capillary and adsorbed water. An existing water retention curve equation is extended for modelling the freezing of capillary water at different void ratios, and a new equation is proposed for the freezing of adsorbed water. The drying branch of the water retention curve was used to combine the Clausius-Clapeyron equation. It is mainly because that the concept of soil freezing process is like that of the drying of unsaturated soils except that ice is substituted by air, air being intruded into the soil pores. Extensive experimental results reported in the literature are used to verify the new model. The prediction by the new model is compared with those of Eq. (1) through (3).

2 Development of a New UWRC Model

2.1 Clausius-Clapeyron Equation

In a closed system, the phase equilibrium between ice and unfrozen water can be described through the Clausius-Clapeyron equation. The phase equilibrium includes mechanical equilibrium (ice and unfrozen water at identical pressure), thermal equilibrium (ice and unfrozen water at identical temperature) and chemical equilibrium (ice and unfrozen water at identical chemical potential). According to Nishimura et al. [8], the Clausius-Clapeyron equation can be expressed as:

$$du_w/p_w - du_i/p_i = \lambda/(T+273.15)dT \tag{4}$$

where u_w and u_i are water and ice pressure, respectively; p_w and p_i are water and ice density, respectively; λ is latent heat of fusion of water (333.7 kJ kg^{-1} at 0 °C) and T is temperature (°C). The integral form of Eq. (4) is:

$$u_w/p_w - u_i/p_i = \int_0^T \lambda/(T + 273.15)dT \tag{5}$$

The value of λ is assumed to be temperature independent and the ice pressure is zero with reference to atmospheric pressure [1]. Equation (5) can be simplified as:

$$u_w = p_w\lambda \ln(T + 273.15)/273.15 \tag{6}$$

It should be noted that the pore water pressure calculated by Eq. (6) is a negative value as the soil temperature is below zero. In frozen soil, the difference between ice pressure and water pressure is defined as the cryogenic suction [11]. As the ice phase is assumed to be at the atmospheric pressure, the cryogenic suction is equal to the negative pore water pressure.

$$s = -u_w \tag{7}$$

Therefore, the relationship between cryogenic suction and soil temperature could be obtained by substituting Eq. (7) into Eq. (6):

$$s = -p_w\lambda \ln(T + 273.15)/273.15 \tag{8}$$

For Eq. (8), the cryogenic suction increases with a decrease in soil temperature. The development of cryogenic suction reduces the potential of pore water and hence the water could retain liquid state in soil pores. In addition, the osmotic suction is ignored in Eq. (8). Similar to unsaturated soils, capillarity and adsorption are two distinct physical mechanisms governing the development of soil cryogenic suction

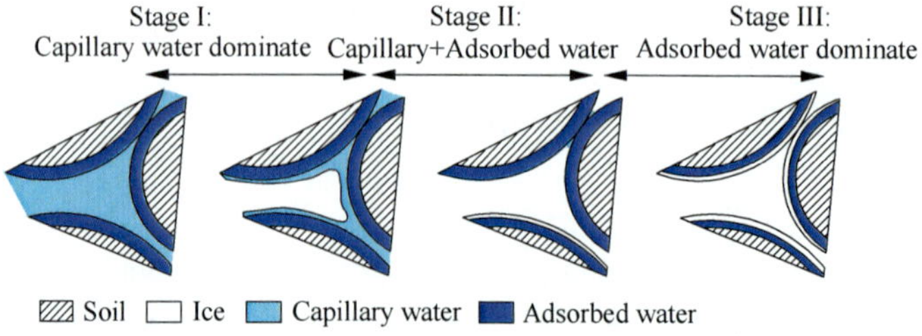

Fig. 1 Idealized freezing process of water in solute-free soil

[1]. The capillarity occurs due to the presence of a curved ice-water interface. The adsorption of water is attributed to the presence of exchangeable cation hydration, mineral surface, or crystal interlayer surface hydration [12].

Figure 1 shows the idealized freezing process of pore water in solute-free saturated soil. The soil specimen consists of soil particle and pore water (capillary and adsorbed water) at the initial state. With a decrease in soil temperature, the phase change firstly occurs in the capillary water which is located relatively far from the soil particle surfaces. The soil pore size distribution, which influences the development of capillary force, is expected to play an important role for the freezing of capillary water. With the temperature further decreases, only a small portion of capillary water which is located in the small pores and the corners of particles retain liquid state. According to the Young-Laplace equation, the water in these small pores could develop high capillary forces and hence retain liquid state at low temperatures. On the other hand, some of the adsorbed water also starts to be frozen in this stage. At the third stage, the phase change occurs in the adsorbed water which surrounds around soil particles. The freezing of absorbed water is influenced by the specific surface area, which control the development of adsorptive forces [17].

2.2 *Freezing of Capillary Water*

The freezing of capillary water, which is governed by capillary force, is dependent on the soil pore size distribution. In this study, the initial void ratio is selected as a parameter to represent the average pore size. Gallipoli et al. [13] proposed an equation to describe the relationship between water content and suction. The equation incorporates the effects of initial void ratio on the volumetric water content at different suctions.

$$\theta_w = [1/(1 + (m_0 e^{m_1} s)^{m_2})]^{m_3} \tag{9}$$

where θ_w is the volumetric water content; e is the initial void ratio; s is the suction and m_0, m_1, m_2 and m_3 are soil parameters. It should be noted that the capillary and adsorbed water were not separated in Eq. (9). In this study, Eq. (9) is modified and adopted to represent the freezing of capillary water as follows:

$$\theta_{cw} = (\theta_s - \theta_{a\max})[1 + (m_0 e^{m_1} s)^{m_2}]^{m_3} \tag{10}$$

where θ_{cw} is the volumetric capillary water content; θ_s is the saturated volumetric water content and θ_{amax} is the maximum volumetric adsorbed water content. The freezing of capillary water could be achieved by substituting Eq. (8) into Eq. (10):

$$\theta_{cw} = (\theta_s - \theta_{a\max})[1/(1 + (m_0 e^{m_1} s)^{m_2})]^{m_3} = (\theta_s - \theta_{a\max}) \cdots\cdots \{1/[1 + (m_0 e^{m_1} p_w \lambda \ln((T+273.15)/273.15))^{m_2}\}^{m_3} \tag{11}$$

Previous studies [7] suggested that the water content at a suction of 1500 k Pa could be used as the residual water content. In this study, the maximum adsorbed water content is determined as the same method as that of residual water content [14]. A suction of 1500 kPa is converted to a temperature about −2 °C based on Eq. (8). Therefore, the value of parameter θ_{amax} could be determined as the unfrozen volumetric water content at a temperature of −2 °C.

2.3 *Freezing of Adsorbed Water*

The freezing of soil pore water at low temperature ranges (< −2 °C) occurs in the adsorbed water. Previous studies [1] showed that the UWRC at the low temperature range has a good agreement with the water retention curve by converting temperature to soil suction through Clausius-Clapeyron equation. Lu [14] proposed an exponential equation to describe the water retention characteristic of absorbed water. In this study, a similar exponential equation is adopted for the freezing of adsorbed water.

$$\theta_{aw} = \theta_{a\max}\{1 - [\exp((T - T_{\min})/T)]^k\} \tag{12}$$

where T_{min} is the temperature that all the pore water is frozen and k is a soil parameter defining the freezing rate of adsorbed water. According to previous study [15], there is a maximum suction about 10^6 kPa corresponding to a zero water content in any porous medium. The maximum matric suction could be converted to a minimum temperature of −259 °C according to Eq. (8). Therefore, the parameter of T_{min} in Eq. (12) is determined as −259 °C.

2.4 *The New UWRC Model*

By combining Eqs. (11) and (12), the total amount of unfrozen water under a given soil temperature could be expressed as:

$$\theta = \theta_{cw} + \theta_{aw} = (\theta_s - \theta_{a\max})\{1/[1 + (m_0 e^{m_1} p_w \lambda \ln((T + 273.17)/273.15))^{m_2}]\}^{m_3} + \theta_{a\max}\{1 - [\exp((T - T_{\min})/T)]^k\} \quad (13)$$

In this equation, five fitting parameters including m_0, m_1, m_2, m_3 and k are included. Furthermore, the parameter k is suggested to be ranged from 0.02 to 0.04, as explained in the "2.4 Freezing of adsorbed water".

3 Verification of the New UWRC Model

3.1 UWRCs of Silt at Different Void Ratios

Azmatch et al. [9] measured the UWRC of Devon Silt at three different initial void ratios. The soil tested is with a specific gravity of 2.65, a clay fraction of 25% and a silt fraction of 75%. Slurry of the soil is prepared and then consolidated under different stresses (50, 100 and 400 kPa). The UWRC is measured with the soil specimens unloaded to a zero stress condition. The initial void ratio of these three specimens is determined as 0.61, 0.58 and 0.55, respectively.

Figure 2 shows the comparisons between the measured and calculated UWRCs of the soil specimen at different initial void ratios. For the measured UWRCs, the unfrozen water content decreases with a decrease in soil temperature. Furthermore, the freezing rate of pore water decreases with a decrease in soil void ratio. During the phase change, more water could be retained liquid state under a given temperature in the soil specimen with a smaller void ratio (0.55). Similar experimental results were observed by Konard [10] showing that the unfrozen water retention capacity increases with decreasing of soil void ratio. It is because that the capillary water in small pores results in high capillary forces than that of the water in large pores

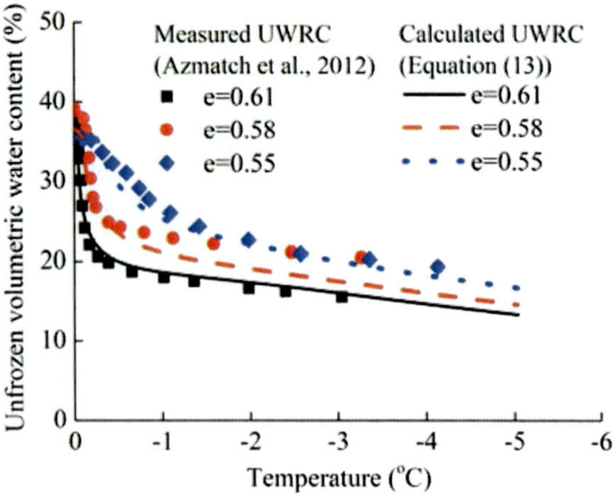

Fig. 2 Comparisons between the measured and calculated UWRCs of silt with different initial void ratios

and hence could retain liquid state at low temperatures. It should be noted that the measured soil UWRCs [9] at void ratios of 0.58 and 0.55 have some cross points when temperatures are smaller than −2 °C. It is probably due to experimental errors, considering the difficulties in the accurate measurement of unfrozen water content. To explore the capability of the new model, Eq. (13) was applied to calculate the UWRC of soil specimen at different initial void ratios. The parameters of m0, m1, m2, m3 and k were obtained by three-dimensional least-square fitting in θ: T: e space with the measured UWRCs at the initial void ratio of 0.61 and 0.55. The same parameters were used to calculate the UWRC at the initial void ratio of 0.58. The proposed model has a high value of R^2 (0.93) for the calibration of model parameters. Moreover, Eq. (13) provides a reasonably good calculation of the UWRC at the initial void ratio of 0.58 through the calibrated parameters. On the other hand, the calculation of unfrozen absorbed water content for the three soil specimens is exactly the same. According to the new model, the increase of unfrozen water retention capacity with decreasing void ratio is mainly contributed by the unfrozen capillary water but not the adsorbed water.

3.2 *UWRC of Clay Over a Wide Temperature Range*

Yoshikawa and Overduin [16] measured the UWRC of Umiat bentonite with different commercial sensors including time domain reflectometry, nuclear magnetic resonance, and frequency domain reflectometry sensor. The measured UWRCs cover a wide temperature range from −50 to 0 °C. In this part, the new model was used to calculate the UWRC of Umiat bentonite over a wide range of temperatures. In addition, as the UWRC under a given initial void ratio was calculated in this section, the parameter of m1 in the new model which represents the effects of void ratio on UWRC was set to zero. On the other hand, three popular existing models found in the literature (Eq. (1) through (3)) were adopted to calculate the UWRCs for comparisons.

Figure 3a shows the comparisons between the measured and calculated UWRCs over a wide temperature range with the new model. The unfrozen capillary and adsorbed water, which are calculated by Eqs. (11) and (12) respectively, are shown in the figure. The aim is to further understand the influence of capillarity and adoption on water freezing. The soil parameter of m_0, m_1, m_2, m_3 and k are 4.62×10^{-3}, 0, 73.577, 0.015 and 0.021 respectively. The parameter of θ_{amax} is determined as 13.5%. The proposed model has a high value of R2 (0.97). The well performance is mainly because the freezing of capillary and adsorbed water is differentiated. The entire freezing process of pore water, which includes both capillarity and adsorption, could be well captured. On the other hand, with soil temperature decreasing from 0 to −2 °C, the unfrozen capillary water content decrease significantly while the unfrozen adsorbed water content almost keeps constant. It shows that the phase change is dominated by capillary water. With the soil temperature further decreasing from −2 to −10 °C, the unfrozen capillary water content continues to decrease but with

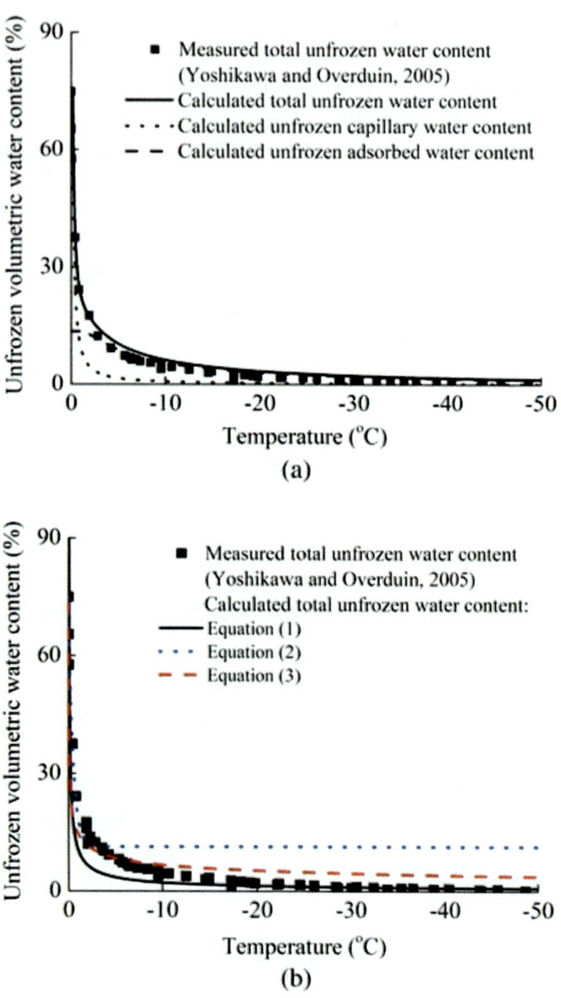

Fig. 3 Comparisons between the measured and calculated UWRCs of clay over a wide range of temperatures: **a** the new model; **b** the existing models

much smaller rates than that in the temperature ranging from −2 to 0 °C. On the other hand, the adsorbed water also starts to be frozen in this stage. It shows that the phase change occurs in both capillary and adsorbed water. With the soil temperature below −10 °C, the unfrozen capillary water content is almost zero and the UWRC is consistent with the freezing curve of adsorbed water. It shows that the phase change dominates by the adsorbed water.

Figure 3b shows the comparisons between the measured and calculated UWRCs with existing models. For the power equation, an overall poor performance was obtained with a low value of R^2 (0.7). The calculated UWRC shows significant discrepancies at the high temperature range (> −10 °C) while a good agreement was obtained with the soil temperature ranging from −50 to −10 °C. On the other hand, the exponential equation also has a low value of R^2 (0.72). The calculated UWRC by exponential equation agrees well with the measured UWRC with soil temperature ranging from −2 to 0 °C. However, the calculated UWRC almost flats with the soil temperature below −2 °C. It is because that the freezing of adsorbed water, which dominates the phase change at the low temperature range, is ignored in the exponential equation. Significant discrepancies were induced between the calculated and measured UWRCs. It can be seen that both the power and exponential equations have poor performance for calculating the UWRC over a wide range of temperatures. For the v-G equation, the value of R^2 is 0.9 which is about 8% smaller than that of the new model.

3.3 *UWRCs of Soil with Different Specific Surface Areas*

Hivon and Sego [17] measured the UWRC of three soil types through the time domain reflectometry method. Each of soil has a different grain size distribution: soil-A, which is a very fine silty sand called Devon silt; a silty sand referred to as soil-B, obtained by mixing 50% of soil-A and 50% of soil-C and a sand referred to as soil C. It should be noted that the specific surface area of soil-A is the largest among these three soils. In addition, as the UWRC under a given initial void ratio was calculated in this section, the parameter of m_1 in the new model which represents the effects of void ratio on UWRC was set to zero. The three popular existing models found in the literature (Eq. (1) through (3)) were also adopted to calculate the UWRC for comparison.

Soil-A: Fine silt sand

Figure 4a shows comparisons between the measured and calculated UWRCs of fine silt sand with the new model. The freezing of capillary water and adsorbed water, which is calculated by Eqs. (11) and (12) respectively, is shown in the figure. The soil parameter of m_0, m_1, m_2, m_3 and k are 3.33×10^{-3}, 0, 63.347, 0.022 and 0.025, respectively. The value of θ_{amax} is determined as 8.89%. It can be seen that the UWRC calculated by the new model has a high value of R^2 (0.97). According to the calculation, the phase change mainly occurs in the capillary water with soil temperature decreasing from 0 to −2 °C. A transition regime which includes the freezing of both capillary and adsorbed water is exhibited with soil temperature further decreasing from −2 to −4 °C. With the soil temperature below −4 °C, the variation of unfrozen volumetric water content is consistent with the freezing curve of adsorbed water. It shows that the phase change only occurs in the absorbed water.

Figure 4b shows the comparisons between the measured and calculated UWRC of fine silt sand with existing models. It can be seen that the power equation shows a low value of R^2 (0.87) indicating a poor performance for calculating the UWRC of fine silt sand. For the exponential equation, the value of R^2 is 0.88 which is similar to that of the power equation. The value of R^2 for power equation and exponential equation is about 11% lower than that of the new model. Furthermore, the exponential equation provides a good calculation with the soil temperature above −2 °C while

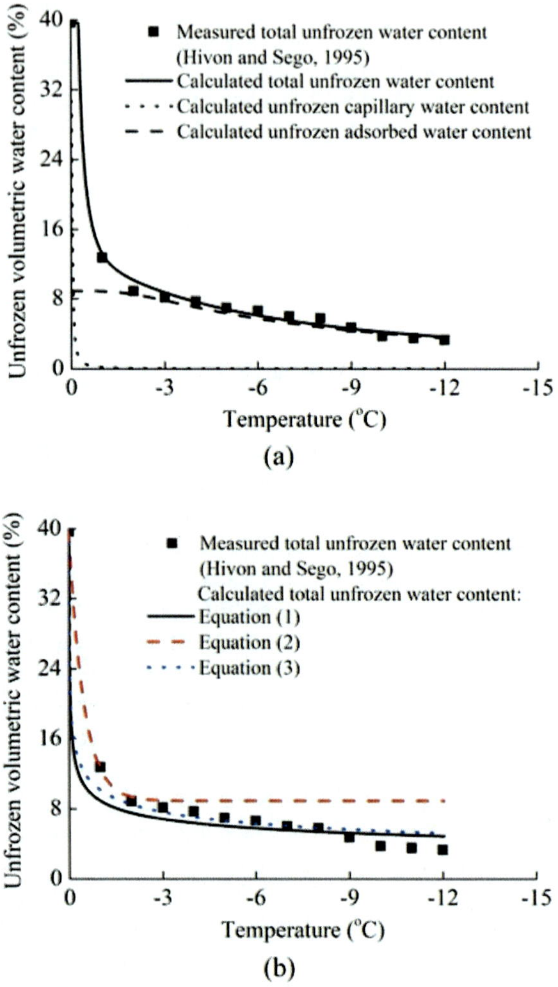

Fig. 4 Comparisons between the measured and calculated UWRCs of fine silt sand: **a** the new model; **b** the existing models

significant discrepancies were exhibited as the soil temperature is below −2 °C. It is mainly because the freezing of adsorbed water is ignored in the exponential equation as explained in the section of "UWRC of clay over a wide temperature range". On the other hand, the v-G equation well calculates the UWRC of fine silt sand, as proved by a relatively high value of R^2 (0.96).

Soil-B: Silt sand

Figure 5a shows comparisons between the measured and calculated UWRC of silt sand with the new model. The soil parameter of m_0, m_1, m_2, m_3 and k are 6.5×10^{-3}, 0, 69.978, 0.018 and 0.026, respectively. The value of θ_{amax} is determined as 3.91%. The value of θ_{amax} for silty sand is 57.14% smaller than that of fine silt sand. It is because that the specific surface area of fine silt sand is larger than that of silt

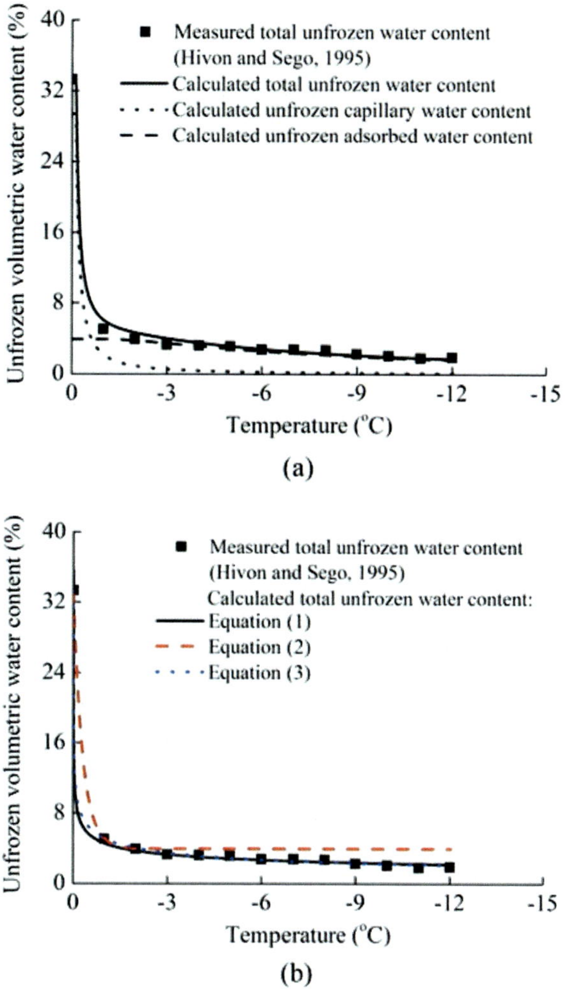

Fig. 5 Comparisons between the measured and calculated UWRCs of silt sand: **a** the new model; **b** the existing models

sand. Furthermore, a high value of R^2 (0.98) is obtained for the calculated UWRC with the new model.

Figure 5b shows the comparisons between the measured and calculated UWRCs of silt sand with existing models. For the power equation, a high value of R^2 (0.98) is obtained. On the other hand, the exponential equation also performs well for calculating the UWRC of silt sand ($R^2 = 0.98$). It is mainly because silt sand has a small specific area and hence a low value of θ_{res} (3.91%). The discrepancy between the measured and calculated UWRCs which is caused by freezing of adsorbed water is insignificant. In addition, v-G equation shows a similar value of R2 (0.98) to that of power and exponential equations. It can be seen that all power equation, exponential equation and v-G equation have good calculations for the UWRC of silt sand which has a relatively small value of specific surface area.

Soil-C: Sand

Figure 6a shows the comparisons between the measured and calculated UWRCs of sand with the new model. As the adsorbed water content is negligible ($\theta_{amax} = 0$) for sand, the parameter of k is equal to zero and the proposed new model is identical as the v-G equation. It shows that the freezing of pore water in sand is only dominated by the capillarity. The soil parameter of m_0, m_1, m_2 and m_3 are equal to 0.79, 0, 73.949 and 0.008, respectively. As expected, a high value of R^2 (1) is achieved. Figure 6b shows the comparisons between the measured and calculated UWRCs of sand with existing models. All power equation, exponential equation and v-G equation have high value of R^2 (1).

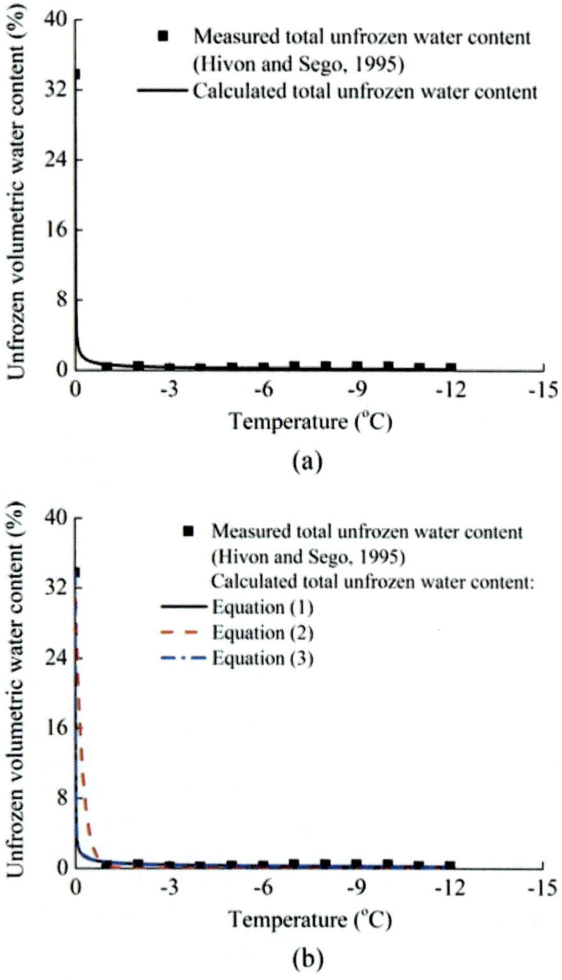

Fig. 6 Comparisons between the measured and calculated UWRCs of sand: **a** the new model; **b** the existing models

The comparisons shown in Figs. 4, 5 and 6 demonstrate that the proposed new model improves the performance for calculating the UWRC of soil with a large specific area such as fine silt sand (see Fig. 4). It is mainly because both capillarity and adsorption contribute to the freezing of pore water in fine silt sand. A better performance was obtained for the new model which explicitly separates the freezing of capillary and adsorbed water. In addition, the proposed new equation has a similar performance with the power equation, exponential equation and v-G equation for calculating the UWRC of the soil with relative small specific surface area such as silt sand (see Fig. 5) and sand (see Fig. 6). For the silt sand and sand, the freezing of pore water is dominated by capillarity.

4 Summary and Conclusions

In this study, a new UWRC model is developed by differentiating the freezing of capillary and adsorbed water. Capillarity and adsorption is considered dominating when the freezing of pore water at higher (above −2 °C) and lower temperatures (below −2 °C), respectively. Moreover, effects of initial void ratio on the freezing of capillary water are explicitly incorporated.

The new model and three existing models are applied to simulate UWRCs of various soils. It is evident that the new model improves the calculation of unfrozen water content under a wide temperature range (from −50 to 0 °C) over existing models. This improvement is attributed to explicit considerations of capillarity and adsorption. Furthermore, only the new model is able to capture effects of initial void ratio on UWRC.

Acknowledgements This work was supported by the National Science Foundation of China (51909205), the Science and Technology Co-ordination and Innovation Project of Shaanxi Province in China (2016KTZDSF03-02) National Key Research and Development Program of China (2017YFD0800501), the innovative talents promotion plan in Shaanxi Province of China (2018KJXX-033), Chinese Postdoctoral Science Foundation (2018M631166) and the Fundamental Research Funds for the Central Universities of China (xjj2018250).

References

1. Lebeau M, Konrad JM (2012) An extension of the capillary and thin film flow model for predicting the hydraulic conductivity of air-free frozen porous media. Water Resour Res 48(7):W07523
2. Sheng DC, Zhang S, Niu FJ, Cheng G (2014) A potential new frost heave mechanism in high-speed railway embankments. Géotechnique 64(2):144–154
3. Zhang S, Sheng DC, Zhao G, Niu FJ, He Z (2015) Analysis of frost heave mechanisms in a high-speed railway embankment. Can Geotech J 53(3):520–529
4. Zhang Y, Michalowski RL (2015) Thermal-hydro-mechanical analysis of frost heave and thaw settlement. J Geotech Geoenviron Eng 141(7):04015027

5. Anderson DM, Tice AR (1972) Predicting unfrozen water contents in frozen soils from surface area measurements. Highway Res Rec 393:12–18
6. Michalowski RL (2015) A constitutive model of saturated soils for frost heave simulations. Cold Reg Sci Technol 22(1):47–63
7. Van Genuchten MT (1980) A closed-form equation for predicting the hydraulic conductivity of unsaturated soils. Soil Sci Soc Am J 44(5):892–898
8. Nishimura S, Gens A, Olivella S, Jardine RJ (2009) THM-coupled finite element analysis of frozen soil: formulation and application. Géotechnique 59(3):159–171
9. Azmatch TF, Sego DC, Arenson LU, Biggar KW (2012) Using soil freezing characteristic curve to estimate the hydraulic conductivity function of partially frozen soils. Cold Reg Sci Technol 83:103–109
10. Konrad JM (1990) Unfrozen water as a function of void ratio in a clayey silt. Cold Reg Sci Technol 18(1):49–55
11. Thomas HR, Cleall PJ, Li YC, Harris C, Kern-Luetschg M (2009) Modelling of cryogenic processes in permafrost and seasonally frozen soils. Géotechnique 59(3):173–184
12. Tuller M, Or D, Dudley LM (1999) Adsorption and capillary condensation in porous media: liquid retention and interfacial configurations in angular pores. Water Resour Res 35(7):1949–1964
13. Gallipoli D, Gens A, Sharma R, Vaunat J (2003) An elasto-plastic model for unsaturated soil incorporating the effects of suction and degree of saturation on mechanical behaviour. Géotechnique 53(1):123–136
14. Lu N (2016) Generalized soil water retention equation for adsorption and capillarity. J Geotech Geoenviron Eng 142(10):04016051
15. Fredlund DG, Xing A (1994) Equations for the soil-water characteristic curve. Can Geotech J 31(4):521–532
16. Yoshikawa K, Overduin PP (2005) Comparing unfrozen water content measurements of frozen soil using recently developed commercial sensors. Cold Reg Sci Technol 42(3):250–256
17. Hivon EG, Sego DC (1995) Strength of frozen saline soils. Can Geotech J 32(2):336–354

Discussion Article on Coastal Engineering-Problems, Research Directions and Needs of Cooperation

Subramaniam Neelamani

Abstract Coastal engineering has become more challenging, since accelerating infrastructure development is happing during the recent years in the coastal area. With increase in global warming, sea level rise is expected to cross 5–8 mm/year in the near future, which will cause significant inundation, sea water intrusion, loss of strength and stability of coastal structures due to increased coastal erosion. It is time to strictly follow the concept of "Reduce, Reuse and Recycle" in all aspects of coastal infrastructure. It is also important to learn and adopt "Defend, Adopt and Retreat" policy appropriately according to the need. This will help for sustainable and holistic socioeconomic development in the coastal area.

Keywords Climate change · Sea level rise · Coastal vulnerability · Adaptation and resilience · Conservation of coastal resources

1 Introduction

Out of 257 countries in the world, 213 countries have coastline (https://www.listofcountriesoftheworld.com/coastline.html). Only 44 countries exist without coastline. The coastline length of Canada is the longest (202,080 km) and the coastline of Tromelin Island is the shortest (3.7 km). All the other countries, the coastline length is in between these two numbers. The total coastline of the world is about 620,000 km. A country with coast is considered as blessed one because of its beneficial ecological services to the people and other life. Compared to interior land, the coastal area attracts more development and attention because of the following reasons:

- The sea is rich in protein and provides many varieties of fish
- Sea is the source of water for desalination and cooling of power plants
- Coastal area is needed for building port, harbor as well as marina, which are needed for international trade

S. Neelamani (✉)
Coastal Management Program, Kuwait Institute for Scientific Research, Kuwait City, Kuwait
e-mail: sneelamani@yahoo.com; nsubram@kisr.edu.kw

A. Garg et al. (eds.), *Proceedings of the 1st Indo-China Research Series in Geotechnical and Geoenvironmental Engineering*, Lecture Notes in Civil Engineering 123,
https://doi.org/10.1007/978-981-33-4324-5_3

- Coastal area is the main attraction for tourism and entertainment
- Coastal area provides many different nonliving resources
- The climate near the coast is moderate due to the presence of cool breeze from sea
- Ocean also acts as sink for treated waste water
- Ocean is the source for future energy and medicines.

The following messages published by The Ocean Conference, United Nations, New York, 5–9 June 2017 are worth noticing:

- More than 600 million people (around 10% of the world's population) live in coastal areas that are less than 10 m above sea level.
- Nearly 2.4 billion people (almost 40% of the population of the world) live within 100 km (60 miles) of the coast.
- The ocean economy, which includes employment, ecosystem and cultural services provided by the ocean, is estimated at between US$ 3–6 trillion/year.
- Over the last two decades, the Greenland and Antarctic ice sheets have been losing mass, glaciers have continued to shrink almost worldwide, and Arctic sea ice and Northern Hemisphere spring snow cover have continued to decrease in extent.
- Over the past three decades, Arctic summer sea ice retreat was unprecedented and sea surface temperatures were anomalously high in at least the last 1,450 years.
- Between 1901 and 2010, global sea level rise increased at an accelerating rate and recent sea level rise appears to have been the fastest in at least during the last 2800 years.
- During the last four decades, 75% of the sea level rise can be attributed to glacier mass loss and ocean thermal expansion. This gives Antarctica alone the potential to contribute more than a meter of sea level rise by 2100 and more than 15 m by 2500.
- Sea level rise leads to coastal erosion, inundations, storm floods, and tidal water encroachment into estuaries and river systems, contamination of freshwater reserves and food crops, loss of nesting beaches, as well as displacement of coastal lowlands and wetlands. In particular, sea level rise poses a significant risk to coastal regions and communities.
- Almost two-thirds of the world's cities with populations of over five million are located in areas at risk of sea level rise.
- The potential costs associated with damage to harbors and ports due to sea level rise could be as high as $ US 111.6 billion by 2050 and $ US 367.2 billion by the end of the century.

2 Main Problems and Challenges

It is essential to understand the varieties of vulnerability issues on the coastal area and how they will affect the people and the infrastructures in the coastal area. How people need to respond to the vulnerability threats on the coast, learn to adopt, build resilience system and protect them so that the services offered by nature in the coastal area can be sustainably utilized by the present as well as future generations. There are thousands of engineering challenges in this area. The basic knowledge in the area of applied mechanics, structural engineering, soil mechanics and foundation, fluid mechanics and dynamics, environmental engineering etc. needs to be upgraded to create cost effective coastal and ocean infrastructure developments. High quality and focused R&D and novel technology development is needed for optimal engineering solutions in this area. A clear understanding of the ocean environment is the primary requirement such as Waves, Tides, Wind, Tsunami, Cyclones, Global warming and Sea level rise problems, Acidification of ocean, Coral reefs, Mangroves and other lives and its interaction for developmental projects. Engineering issues persists in areas such as Assessment of environmental forces on different marine structures, Sediment transport problems, Marine pollution, field measurements of Metocean parameters and sea survey, prevention of corrosion on structures etc. Optimizing the design is always a challenge in many areas such as Subsea tunnels and pipelines, Offshore oil and gas exploitation, Coastal erosion control, artificial beach developments, Port and Harbor structures, Sea water intake Structures, Artificial Island Developments etc. Challenges and opportunity exists for field measurements and ocean observation, Force reduction techniques on marine structures, development of new materials for innovative marine structures, Aquaculture, Desalination for potable water, Environmental impact due to marine space development, Shipping & Navigation issues, Dredging/Reclamation, Disposal of wastes in the marine environment, Renewable energy extraction from the Ocean (Offshore Wind Energy/Wave Energy/Tidal Energy/Ocean thermal Energy/Ocean Currents), Prediction of design conditions, Environment impact of new and existing projects, Issues in the Integrated coastal zone management for sustainable development, Innovative new structures for Economy and Improved life span, Remote Sensing Applications, managing the overexploitation of Living and Non-Living resources, Tourism and development, Sea sports, Development of New Township along the Coastline, Coastguard Establishments for security. The list mentioned above is not exhaustive. Each and every activity in the coastal and ocean space results in significant social and economic benefits and environmental impacts, technical challenges and financial implications. Some of the challenges and probable solutions will be revealed during the lecture for the benefit of students. Human Resource Development is very vital.

3 Research Directions and Need for Cooperation

The main research directions in coastal and ocean engineering need to focus on innovative marine structures for cost reduction, ease of construction and installation, less negative impact on environment, new materials from corrosion reduction, hydrodynamic force reduction, less risk of failures of different marine structures etc. Future Sea level rise prediction is another area with high degree of uncertainty. Vulnerability and risk assessment of the coast due to sea level rise, Tsunami, cyclone and storm surge is another area for focused research. Educating the coastal community to lead a holistic life with nature understanding the nature of coastal and ocean environment is another important area. Most of the coastal and ocean engineering issues are global. Hence, cooperation between different countries in these areas will yield results which will help for the significant holistic global development.

A Comprehensive Study for Assessing Parameters Influencing Tensile Strength Behaviour of Fine-Grained and Coarse-Grained Soils

Peddireddy Sreekanth Reddy, He Huang, Xilong Huang, Yusuf Erzin, Mei Guixiong, Ankit Garg, and Bendadi Hanumantha Rao

Abstract Development of tensile cracks in earthen structures is majorly controlled by tensile strength, which is affected by a number of parameters. The influence of different soil properties and their relative dominance on tensile strength characteristics of both fine- and coarse-grained soils has been brought out by systematic investigations using Artificial Neural Network (ANN) technique, in the present study. A few important soil parameters considered are degree of saturation, dry density, porosity, suction, plasticity index, fines content, relative density, average particle diameter, and sand content. Several ANN models, independently for fine- and coarse-grained soils, consisting of minimum one to a maximum six input parameters have been developed. Results demonstrate that influence of parameter(s) is distinct on tensile strength of fine- and coarse-grained soils. Results also clearly manifested usefulness of ANN tool to discern which parameter(s) could largely govern the tensile behaviour

P. S. Reddy
Department of Civil Engineering, NIT Mizoram, Aizawl 796012, India
e-mail: psreddy.civil.phd@nitmz.ac.in

H. Huang · X. Huang · A. Garg
Guangdong Engineering Center for Structure Safety and Health Monitoring, Shantou University, Shantou, China
e-mail: 18hhuang2@stu.edu.cn

X. Huang
e-mail: 18xlhuang1@stu.edu.cn

A. Garg
e-mail: ankit@stu.edu.cn

Y. Erzin
Department of Civil Engineering, Celal Bayar University, Muradiye, Manisa, Turkey
e-mail: yusuf.erzin@cbu.edu.tr

M. Guixiong
School of Civil Engineering and Architecture, Guangxi University, Guangxi, China
e-mail: meiguox@163.com

B. H. Rao (✉)
School of Infrastructure, IIT Bhubaneswar, Khordha 752050, Odisha, Bhubaneswar, India
e-mail: bhrao@iitbbs.ac.in

A. Garg et al. (eds.), *Proceedings of the 1st Indo-China Research Series in Geotechnical and Geoenvironmental Engineering*, Lecture Notes in Civil Engineering 123,
https://doi.org/10.1007/978-981-33-4324-5_4

of soils. Amongst many parameters, fines content, sand content and average particle diameter are found prominently influencing and ascribed as must to include parameters for determining the tensile strength of soils respectively. It has been noticed that the significance of these grain size parameters is analogous to microstructural hydromechanics of soil and, thereby, accounted for universal tensile behaviour of soils.

Keywords Tensile behaviour · Grain size characteristics · ANN modelling · Fine-grained soils · Coarse-grained soils

1 Introduction

Generally, cohesive soils such as clays do exhibit significant tensile strength (σ_t), whereas non-cohesive soils like sands barely show any tensile strength [1]. In comparison, the tensile strength is relatively small in magnitude with its compressive or shear strength of a soil. It is important to understand the tensile behaviour of soils owing to its strong correlation with cracking characteristics of a soil, specifically in embankments, dams, slopes, retaining walls, pavements, river banks, hydraulic barriers, capping systems of waste containment facilities, soil erosion control systems, etc. [1–6]. Furthermore, cracking and crumbling are a common phenomenon in fine-grained soils, and they are highly sensitive to environmental changes such as moisture, temperature, and compaction state. These phenomena are also strongly dependent on tensile strength of a soil. Thus, recognizing the probability of tensile crack development becomes one of the main applications of determining the tensile strength of a soil [5, 7]. Moreover, knowledge on tensile behavior is vital in assessing the tensile zone widening in earthen structures, apart from deriving its help in ascertaining which failure criteria are more general or what forces between the individual grains affect the soil structure arrangements [8–11].

Many researchers reported that tensile strength of soils, which include both fine-grained and coarse-grained, depends on several soil properties such as degree of saturation, dry density, clay content (and its type), fines content, void ratio, consistency limits, suction, amount of organic matter, exchangeable cations, drying and wetting cycle, etc. [3, 12–20]. Numerous empirical correlations relating σ_t with one or more than one soil property have been proposed by previous researchers. However, most of the relationships postulated were developed either belonging to a specific soil(s) or considering only a single parameter (viz., suction, plasticity index, liquid limit, CEC, clay content, water content) [5, 14, 15, 18, 21–24]. Venkataramana et al. [25] have summarized the various relationships available in the literature correlating tensile strength with the soil property. The authors made a remark that these relationships cannot be regarded as generalized relationships. Moreover, most of the equations are best suited to indicate that the correlations can exist between tensile strength and the corresponding soil property. On the other hand, studies also report that there is an inherent advantage in developing generalized relationships based on multiple soil properties, as it is obligatory to represent the overall behavior of a particular soil through these generalized relationships [25]. Incidentally, the proposed equations, which mostly require a single input parameter, also failed to postulate which parameter has had greater influence on the tensile behavior of soils.

Artificial Neural Network, ANN, one of artificial intelligence techniques, is a diagnostic procedure to imitate the behavior of brain and human nervous system [26]. ANNs are very sophisticated modeling techniques, capable of modeling extremely complex functions [27]. Therefore, ANNs, with their remarkable ability to derive a general solution from complicated and imprecise data, can be used to extract patterns and detect trends that are too complex to be noticed by either humans or other computer techniques [28]. Perhaps ANN has regarded as the best meritorious tool to delineate critically the influence of individual parameter on the overall behavior.

The objective of the present study is to develop models with specific focus on assessing the influence of parameter(s) either individually or combined with one or more parameters to determine the tensile strength of fine- and coarse-grained soils. Additionally, efforts are also devoted to carry out the parametric sensitivity analysis, which would help in identifying the critical parameter(s) that is vitally important for inducing the tensile strength in soils. For achieving the objective of the study, ANN tool has been adopted. With the understanding of sensitivity analysis on parameters for predicting tensile strength of soil, designing, estimating, and controlling for tensile cracking in earth engineering can be accomplished, apart from enhancing further quantified researches on tensile strength of soils and reinforced soils. Such study can also aid in improving fundamentals behind the development of tensile strength in soils (fine-grained and coarse-grained).

2 Artificial Neural Networks

ANN approach could be applied to interpret the connection between soil behavior and its various properties, as there is no need for assumptions or initial-condition requirements for modelling. There were many successful applications of ANN tool in geotechnical engineering to solve a simple to very complex problems [29–31].

A neural network consists of an interconnected group of artificial neurons. In the ANN network, there is one input layer, a hidden layer(s) and one output layer. Each layer may have different number of neurons, and even a different transfer function. The weighted sum of inputs is transferred to the hidden neurons, where it is transformed using an activation function. The common transfer function widely used in the literature is the logistic sigmoid function. The outputs of hidden neurons, in turn, act as inputs to the output neuron where it undergoes another transformation. The output of a feed-forward neural network with two hidden layers and one output neural network is given by

$$Output = f^3(W^3 f^2(W^2 f^1(W^1 p + b^1) + b^2) + b^3) \quad (1)$$

where, W^1 is the weight matrix for the first layer, W^2 is the weight matrix for the second layer, W^3 is the weight matrix of third layer, f^1 is the hidden neuron activation function, f^2 is the output neuron activation function, f^3 is the output hidden neuron activation, b^1 is the bias vector first layer, b^2 is the bias vector second layer and b^3 is the bias vector third layer.

For developing model, network training is carried out based on the data. Network training is a process by which the connection weights and biases of the ANN are adapted through a continuous process of simulation by the environment in which the network is embedded. The primary goal of training is to minimize an error function by searching for a set of connection strengths and biases that cause the ANN to produce outputs that are equal or close to targets. In other words, training aims at estimating the parameters (W^1, W^2, b^1, and b^2) by minimizing an error function such as the mean square error of the output values expressed as:

$$MeanSquareError = \sum_{i=1}^{N} \frac{t_i - o_i}{N} \tag{2}$$

where, N is the number of data points, t_i is the measured value, and o_i is the predicted value. The minimization procedure relies on a numerical optimization of a non-linear objective function.

3 Procedure of Developing ANN Models for Predicting Tensile Strength

3.1 Fine-Grained Soils

Ten different ANN models designated as ANNF-0 to ANNF-9 (refer to Table 1) were developed. It is appreciated that these models are not only help in predicting the tensile strength of fine-grained soils based on the employed soil property, but also aid in examining the influence of latter property on the tensile behavior of former soil. For this purpose, the data related to tensile strength reported by several studies [4, 6, 14, 21, 22, 32–42] for different types of soils were collected and used. Fines content (FC) (i.e. grain size characteristic), porosity (n), dry density (ρ_{dry}), plasticity index (PI) (representing consistency limits), suction (ψ), water content (w), and degree of saturation (S_r) were selected as input parameters for fine-grained soils for developing ANN models. Tables 1 and 2 present the information of parameters, number of data sets, and input &output parameters used in the ANN model development.

Table 1 Parameters considered for developing ANN models along with data sets for fine-grained soils

Model designation	Number of parameters	Parameter(s) designation						Number of data sets
ANNF-0	Two	ρ_{dry}	w					217
ANNF-1	One	Ψ	–	–	–	–	–	78
ANNF-2	Two	Ψ	S_r	–	–	–	–	78
ANNF-3	Two	PI	FC	–	–	–	–	217
ANNF-4	Three	PI	FC	ρ_{dry}	–	–	–	217
ANNF-5	Four	PI	S_r	ρ_{dry}	FC	–	–	217
ANNF-6	Three	Ψ	FC	–	–	–	–	217
ANNF-7	Five	PI	n	S_r	ρ_{dry}	FC	–	192
ANNF-8	Five	Ψ	PI	n	ρ_{dry}	FC	–	154
ANNF-9	Six	Ψ	S_r	PI	n	FC	ρ_{dry}	154

3.2 Coarse-Grained Soils

Twelve different ANN models designated as ANNC-0 to ANNC-11 (refer to Table 3) were developed for the prediction of tensile strength of coarse-grained soils and also to examine the influence of soil property on it. For this purpose, the data reported by Ibarra et al. [24], Kim and Hwang [43], Lu et al. [44], Araki et al. [45], Zhang et al. [34], Farrell et al. [46], Jindal et al. [47], Kim et al. [48] and Lu et al. [49] for different types of soils were collected and used. The parameters such as sand content (SC), average particle diameter (D_{50}), degree of saturation (S_r), porosity (n), dry density (ρ_{dry}) and relative density (D_r) were selected as input parameters for developing ANN models. Tables 3 and 4 present the information of parameters, number of data sets, and input & output parameters used in the ANN model development.

3.3 ANN Models

The collected data from the literature was then divided into three subsets, i.e. a training, a testing, and a validation set. The training set is assigned to determine the connection weights of the neural network model. An independent testing set is aimed to measure the performance of the model after training. Whereas, the validation set is used to verify the result of model after training and testing. In this study, 80% of the data for training, 10% for testing of each model, and the remaining 10% for validation purpose were used. Prior to such distribution, numerous trials (not shown in the results) were carried out considering the time taken to compute the accuracy of each model. To maximize network speed and accuracy, it is necessary to normalize the input and output data. The input-output data used for each ANN model were

Table 2 Details of input and output parameters employed in ANN models developed for fine-grained soil

ANN model	Parameters used	Minimum value	Maximum value	Mean value	Standard deviation
ANNF-0	ρ_{dry}	0.840	2.082	1.537	0.262
	w	0.000	27.790	7.445	7.288
	σ_t	0.400	14,875.000	308.670	1332.420
ANNF-1	Ψ	0.000	100,000.000	8714.500	20,058.100
	σ_t	0.790	466.000	189.592	174.215
ANNF-2	Ψ	0.000	100,000.000	8714.500	20,058.100
	S_r	16.200	100.000	77.998	21.176
	σ_t	0.790	466.000	189.592	174.215
ANNF-3	PI	8.900	268.000	33.223	43.478
	FC	60.600	100.000	84.716	15.909
	σ_t	0.400	14,875.000	308.670	1332.420
ANNF-4	PI	8.900	268.000	33.223	43.478
	FC	60.600	100.000	84.716	15.909
	ρ_{dry}	0.840	2.082	1.537	0.262
	σ_t	0.400	14,875.000	308.670	1332.420
ANNF-5	PI	8.900	268.000	33.223	43.478
	S_r	14.316	116.880	74.618	22.601
	ρ_{dry}	0.840	2.082	1.537	0.262
	FC	60.600	100.000	84.716	15.909
	σ_t	0.400	14,875.000	308.670	1332.420
ANNF-6	Ψ	0.100	100,000.000	7150.000	18,374.800
	FC	20.000	100.000	86.025	17.601
	σ_t	0.000	466.000	166.992	164.095
ANNF-7	PI	8.900	268.000	33.223	43.478
	FC	60.600	100.000	84.716	15.909
	S_r	14.316	116.880	74.618	22.601
	n	0.208	0.678	0.419	0.089
	ρ_{dry}	0.840	2.082	1.537	0.262
	σ_t	0.400	14,875.000	308.670	1332.420
ANNF-8	Ψ	14.316	100.000	74.645	23.306
	PI	8.900	268.000	37.577	49.271
	n	0.208	0.678	0.414	0.096
	FC	64.000	100.000	89.798	12.572
	ρ_{dry}	0.847	2.082	1.576	0.263
	σ_t	3.480	14,875.000	407.930	1525.520

(continued)

Table 2 (continued)

ANN model	Parameters used	Minimum value	Maximum value	Mean value	Standard deviation
ANNF-9	Ψ	0.000	100,000.000	11,004.100	22,334.600
	S_r	16.200	100.000	75.449	22.600
	PI	23.500	268.000	55.683	70.623
	n	0.281	0.680	0.381	0.085
	FC	70.000	100.000	92.678	10.451
	ρ_{dry}	0.840	1.899	1.641	0.230
	σ_t	0.790	466.000	240.800	165.798

Table 3 Parameters considered for developing ANN models along with data sets for coarse-grained soils

Model designation	Number of parameters	Parameter(s) designation					Number of data sets
ANNC-0	Two	ρ_{dry}	w				234
ANNC-1	One	D_r	–	–	–		86
ANNC-2	Two	D_r	S_r	–	–	–	86
ANNC-3	Three	n	D_r	S_r	–	–	86
ANNC-4	Four	n	D_r	SC	S_r	–	86
ANNC-5	Four	D_{50}	n	D_r	S_r	–	86
ANNC-6	Five	SC	n	D_{50}	D_r	S_r	86
ANNC-7	Three	ρ_{dry}	S_r	n	–	–	234
ANNC-8	Three	n	ρ_{dry}	D_{50}	–	–	234
ANNC-9	Four	n	S_r	ρ_{dry}	D_{50}	–	234
ANNC-10	Four	n	S_r	SC	ρ_{dry}	–	234
ANNC-11	Five	n	S_r	SC	ρ_{dry}	D_{50}	234

scaled to lie between −0.9 and 0.9, by using Eq. (3).

$$x_{norm} = 1.8 * \frac{(x - x_{\min})}{(x_{\max} - x_{\min})} - 0.9 \quad (3)$$

where, x_{norm} is the normalized value, x is the actual value, $x_{\max}$ is the maximum value and x_{min} is the minimum value of the collected data.

In the current study, a feed-forward network with a back-propagation algorithm, most commonly used neural network structure [50], was used to build the ANN models. While training and testing of feed-forward networks, the neural network toolbox of MATLAB 7.0 was used [51]. The Levenberg-Marquardt back-propagation learning algorithm [52], was used in the training stage. First, one hidden layer was

Table 4 Details of input and output parameters employed in the ANN model developed for coarse-grained soil

ANN model	Parameters used	Minimum value	Maximum value	Mean value	Standard deviation
ANNC-0	ρ_{dry}	0.840	2.082	1.537	0.262
	w	0.000	27.790	7.445	7.288
	σ_t	0.000	422.300	28.680	86.220
ANNC-1	D_r	26.000	72.000	49.478	15.662
	σ_t	0.410	422.300	66.093	132.171
ANNC-2	D_r	26.000	99.000	60.000	24.711
	S_r	1.730	74.733	13.200	14.842
	σ_t	0.410	422.300	66.093	132.171
ANNC-3	N	0.250	0.421	0.364	0.058
	D_r	26.000	99.000	60.000	24.711
	S_r	1.730	74.733	13.200	14.842
	σ_t	0.410	422.300	66.093	132.171
ANNC-4	N	0.250	0.421	0.364	0.058
	D_r	26.000	99.000	60.000	24.711
	SC	55.900	100.000	89.490	18.014
	S_r	1.730	74.733	13.200	14.842
	σ_t	0.410	422.300	66.093	132.171
ANNC-5	D_{50}	0.220	0.269	0.235	0.019
	N	0.250	0.421	0.364	0.058
	D_r	26.000	99.000	60.000	24.711
	S_r	1.730	74.733	13.200	14.842
	σ_t	0.410	422.300	66.093	132.171
ANNC-6	SC	55.900	100.000	89.490	18.014
	n	0.250	0.421	0.364	0.058
	D_{50}	0.220	0.269	0.235	0.019
	D_r	26.000	99.000	60.000	24.711
	S_r	1.730	74.733	13.200	14.842
	σ_t	0.410	422.300	66.093	132.171
ANNC-7	ρ_{dry}	0.913	2.086	1.667	0.253
	n	0.250	0.662	0.408	0.070
	S_r	0.000	100.000	30.236	26.123
	σ_t	0.000	422.300	28.680	86.220
ANNC-8	ρ_{dry}	0.913	2.086	1.667	0.253
	n	0.250	0.662	0.408	0.070
	D_{50}	0.100	0.451	0.269	0.104

(continued)

Table 4 (continued)

ANN model	Parameters used	Minimum value	Maximum value	Mean value	Standard deviation
	σ_t	0.000	422.300	28.680	86.220
ANNC-9	ρ_{dry}	0.913	2.086	1.667	0.253
	n	0.250	0.662	0.408	0.070
	S_r	0.000	100.000	30.236	26.123
	D_{50}	0.100	0.451	0.269	0.104
	σ_t	0.000	422.300	28.680	86.220
ANNC-10	ρ_{dry}	0.913	2.086	1.667	0.253
	n	0.250	0.662	0.408	0.070
	S_r	0.000	100.000	30.236	26.123
	SC	55.000	100.000	86.787	16.591
	σ_t	0.000	422.300	28.680	86.220
ANNC-11	ρ_{dry}	0.913	2.086	1.667	0.253
	n	0.250	0.662	0.408	0.070
	S_r	0.000	100.000	30.236	26.123
	SC	55.000	100.000	86.787	16.591
	D_{50}	0.100	0.451	0.269	0.104
	σ_t	0.000	422.300	28.680	86.220

selected and then increased to two, if needed, for obtaining the best performance. While for determining the optimum number of neurons in the hidden layer of a model, the number of neurons was first selected as 1 and then increased in steps of 1 with each additional time. Different transfer (activation) functions such as log-sigmoid and tan-sigmoid were used to achieve the best performance in training as well as in testing. Two momentum factors, were selected (i.e. 0.01 and 0.001), separately, for the training process to explore the most efficient ANN architecture.

As mentioned earlier, over-fitting gives rise to multilayer feed-forward networks to memorize training patterns and so the networks cannot acquaint well to new data. Therefore, training started with a small number of epochs of 10 and continued incrementing by 10 till the onset of specialized training as reflected in the reversal of the downward trend of the error for testing data. The maximum number of epochs to training was selected as 500. Finally, the ANN model output (predicted tensile strength) was compared with the actual measured tensile strength using the performance indices, namely, the coefficient of determination, R^2, given by Eq. (4), and the root mean square error, *RMSE*, given by Eq. (5), to check the performance of each ANN model developed in this study.

$$R^2 = 1 - \frac{\sum_{i=1}^{N}\left(y_i - \hat{y}_i\right)^2}{\sum_{i=1}^{N}\left(y_i - \overline{y}_i\right)^2} \quad (4)$$

$$RMSE = \sqrt{\frac{1}{N}\sum_{i=1}^{N}\left(y_i - \hat{y}_i\right)^2} \quad (5)$$

where, y is the measured value; ŷ is the predicted value; $\bar{y}$ is the mean of the measured data; and N is the number of samples. Three kinds of *RMSE* values (i.e. for training error, testing error and validation error) were calculated during the performance of ANN models.

In order to determine the optimum network geometry, performance of network during the training, testing and validation stages was evaluated for each network size until no significant improvement occurred. Details of optimal performance of the networks are given in Tables 5 and 6 for each developed ANN model of fine- and coarse-grained soils respectively.

4 Results

Independent ANN analysis was carried out for fine- and coarse-grained soils. Modelling was initially started with one input parameter and progressively advanced by keep adding more number of parameters to a maximum of six for fine-grained and five for coarse-grained soils were in built in models (refer to Tables 1 and 3). In total, 10 ANN models for fine-grained soils and 12 ANN models for coarse-grained soils were developed for predicting their tensile behaviour (refer to Tables 1, 3, 5 and 6). In ANN modeling, though all of the parameters are treated as independent, they, in fact, might exhibit interdependence influence among themselves. For example: suction generally increases with decrease in water content. Similarly, relative density reflects the information of porosity and dry density. Thus, with such assumption suitable selection of parameters in each ANN model can help to interpret the influence and contribution of an index on tensile strength of soils. Among all the parameters, saturation (or water content) and dry density are considered as external factors because they can be controlled artificially for a specific soil sample. The remaining parameters such as fines content, suction, porosity, and plasticity index of fine-grained soils, and sand content, porosity, and D_{50} of coarse-grained soils are considered as internal factors (soil type dependent).

Figure 1a–i shows the performance of ANN models developed for predicting tensile strength of fine-grained soils based on the designated models listed in Tables 1 and 2. It is seen from the figures that for most of the models, measured σ_t has a good performance agreement with predicted σ_t, which indicates that ANN developed some reliable models to meet various kinds of soil conditions. The coefficients of determination obtained from the results presented in Fig. 1a–i are summarized in Table 5. Other details describing the performance about the models are also presented in the same table as an additional information. Among ten models, three models namely ANNF-0 (considers ρ_{dry} and w), ANNF-1 (involves ψ only parameter) and

Table 5 Details of the most suitable model of fine-grained soils

Model no.	Suitable model	Training performance	Test performance	Validation performance	Training error	Testing error	Validation error	R^2	Algorithm	Error function	Hidden activation	Output activation
ANNF-0	Fail											
ANNF-1	RBF 1-27-1	0.917929	0.970546	0.875435	2090.568	1156.023	596.641	0.8529	RBFT	SOS	Gaussian	Identity
ANNF-2	MLP 2-8-1	0.904479	0.911092	0.911784	3607.163	2531.953	1498.277	0.9194	BFGS 295	SOS	Tanh	Tanh
ANNF-3	RBF 2-30-1	0.837764	0.983809	0.844451	110858.5	89782.39	2631750	0.6647	RBFT	SOS	Gaussian	Identity
ANNF-4	MLP 3-9-1	0.996664	0.98056	0.986194	7752.994	5932.93	3784.438	0.9926	BFGS 210	SOS	Tanh	Exponential
ANNF-5	MLP 4-8-1	0.996911	0.983196	0.997195	7176.065	5440.401	1236.217	0.9934	BFGS 59	SOS	Tanh	Exponential
ANNF-6	MLP 2-6-1	0.907736	0.968157	0.995075	2998.212	1167.751	817.1946	0.9083	BFGS 57	SOS	Logistic	Logistic
ANNF-7	MLP 5-9-1	0.995079	0.998945	0.885349	5640.715	830911.8	3540.453	0.9873	BFGS 56	SOS	Tanh	Logistic
ANNF-8	MLP 5-11-1	0.999383	0.993802	0.980745	1744.904	1447.135	319.6047	0.9986	BFGS 278	SOS	Tanh	Exponential
ANNF-9	MLP 6-9-1	0.96416	0.999819	0.999757	881.269	8.31856	131.3318	0.9462	BFGS 99	SOS	Logistic	Logistic

RBF: Radial Basis Function; MLP: multilayer perceptron; RBFT: Redundant Byzantine Fault Tolerance; BFGS: Broyden–Fletcher–Goldfarb–Shan

Table 6 Details of most suitable model of coarse-grained soils

Model no.	Suitable model	Training performance	Test performance	Validation performance	Training error	Test error	Validation error	R^2	Algorithm	Error function	Hidden activation	Output activation
ANNC-0	Fail											
ANNC-1	Fail											
ANNC-2	RBF 2-19-1	0.966993	0.996557	0.945282	600.0347	88.75335	0.761291	0.9425	RBFT	SOS	Gaussian	Identity
ANNC-3	RBF 3-15-1	0.931401	0.999106	0.916926	1223.366	160.5643	358.3348	0.9024	RBFT	SOS	Gaussian	Identity
ANNC-4	MLP 4-5-1	0.972285	0.999971	0.869041	506.4713	1.593397	2	0.9524	BFGS 62	SOS	Tanh	Identity
ANNC-5	MLP 4-5-1	0.972267	0.999922	0.834635	505.0896	2.444669	0.685654	0.9523	BFGS 50	SOS	Logistic	Identity
ANNC-6	MLP 5-8-1	0.978085	0.986267	0.999977	366.9326	211.0989	7.43706	0.9632	BFGS 72	SOS	Tanh	Exponential
ANNC-7	MLP 3-7-1	0.968398	0.984371	0.963098	272.6504	73.43076	0.861709	0.9401	BFGS 130	SOS	Logistic	Logistic
ANNC-8	MLP 3-5-1	0.95385	0.98692	0.926223	393.1459	39.93167	1.347998	0.9142	BFGS 193	SOS	Logistic	Logistic
ANNC-9	MLP 4-9-1	0.959318	0.957935	0.992753	291.8985	288.6391	71.72475	0.9272	BFGS 85	SOS	Tanh	Tanh
ANNC-10	MLP 4-6-1	0.963731	0.953074	0.996227	261.1136	348.3972	37.55846	0.9335	BFGS 67	SOS	Tanh	Tanh
ANNC-11	MLP 5-4-1	0.972986	0.956662	0.995824	195.2034	312.3867	41.34993	0.9484	BFGS	SOS	Logistic	Tanh

RBF: Radial Basis Function; MLP: multilayer perceptron; RBFT: Redundant Byzantine Fault Tolerance; BFGS: Broyden–Fletcher–Goldfarb–Shann

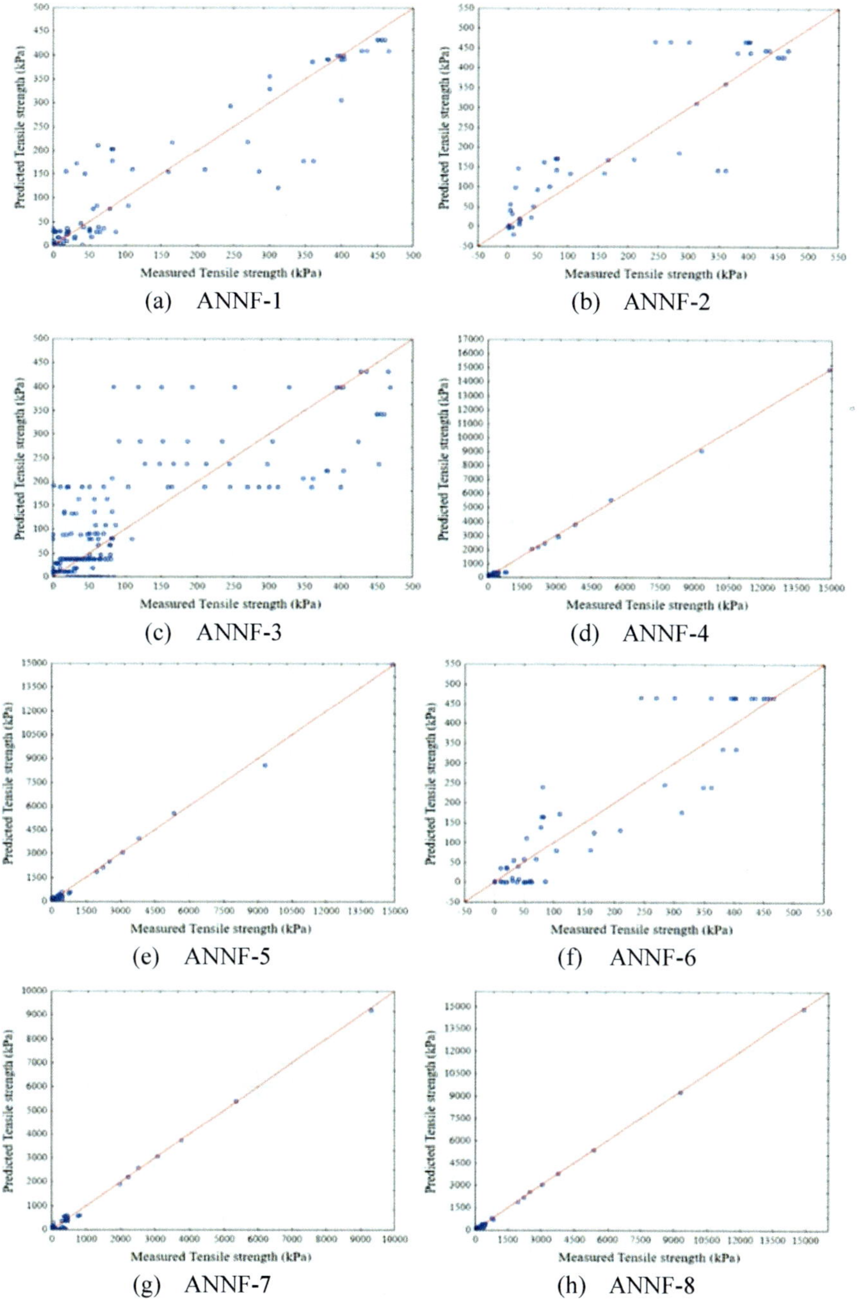

Fig. 1 Measured versus predicted tensile strength of fine-grained soils using the ANNF-1 to ANNF-9 models

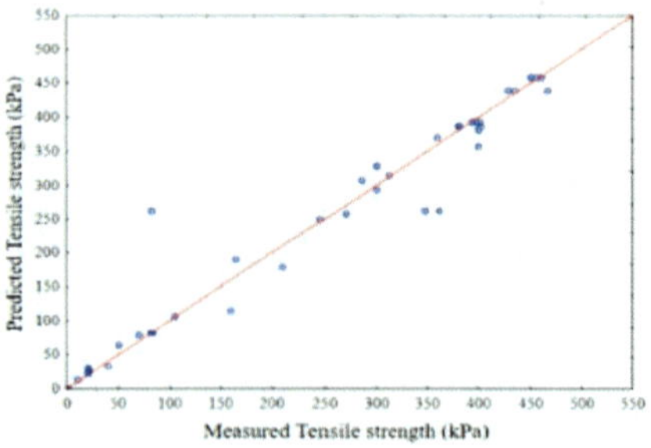

(i) ANNF-9

Fig. 1 (continued)

ANNF-3 (includes *PI* and *FC*) have failed to build relationships between parameters and dependent σ_t. This may be attributed to alone consideration of external factors or internal factorsfor predicting σ_t and as such, ignored the fundamental functions of internal factors and direct effect of external factors. Remaining models, which employed both internal and external factors with more numbers of parameters, exhibited good relationships with tensile strength, as it is obvious that the coefficient of determination (R^2) is more than 0.9 in all of these cases.

Figure 2a–j depict the performance of ANN models programmed for predicting tensile strength of coarse-grained soils, according to selected models from Table 3 and 4. The results pertaining to ANNC-0 (focuses on ρ_{dry} and w) and ANNC-1 (cares about D_r only) were not included as they failed to perform as desired due to the limitation in the parameters selected. It can be observed from the figures that ANNC-2 to ANNC-11 models perform very well with a good correlation between soil parameters and target variable, markedly the value of coefficient of correlation (i.e. R^2) is greater than 0.9. More specific details about the performance of these models are summarized in Table 6.

Based on the results presented herein, two important observations can be made. First, many parameters can influence the tensile behavior of fine- and coarse-grained soils and the second is the effect of parameters is distinct on fine- and coarse grained soils. Evidently, these results do not facilitate for inference of which parameter(s) is crucial or otherwise largely govern the tensile behavior of these soils. It is appreciated that such knowledge is critical for reliable determination of the tensile strength, which in turn assists accurate estimation of soil engineering properties such as cracking, crumbling, etc. With this in view, sensitivity analyses were further carried out to examine the relative significance of each input parameter on the tensile strength prediction. For this purpose, a simple and innovative technique proposed by Garson [53] and verified by Shahin et al. [26] was used to interpret the relative dominance of each input parameter on σ_t. The technique, prima facie, works on the basis of examining the connection weights of the trained network. For a network with one hidden layer, the technique requires a process of partitioning the hidden output connection weights into components associated with each input node [26]. Therefore, based

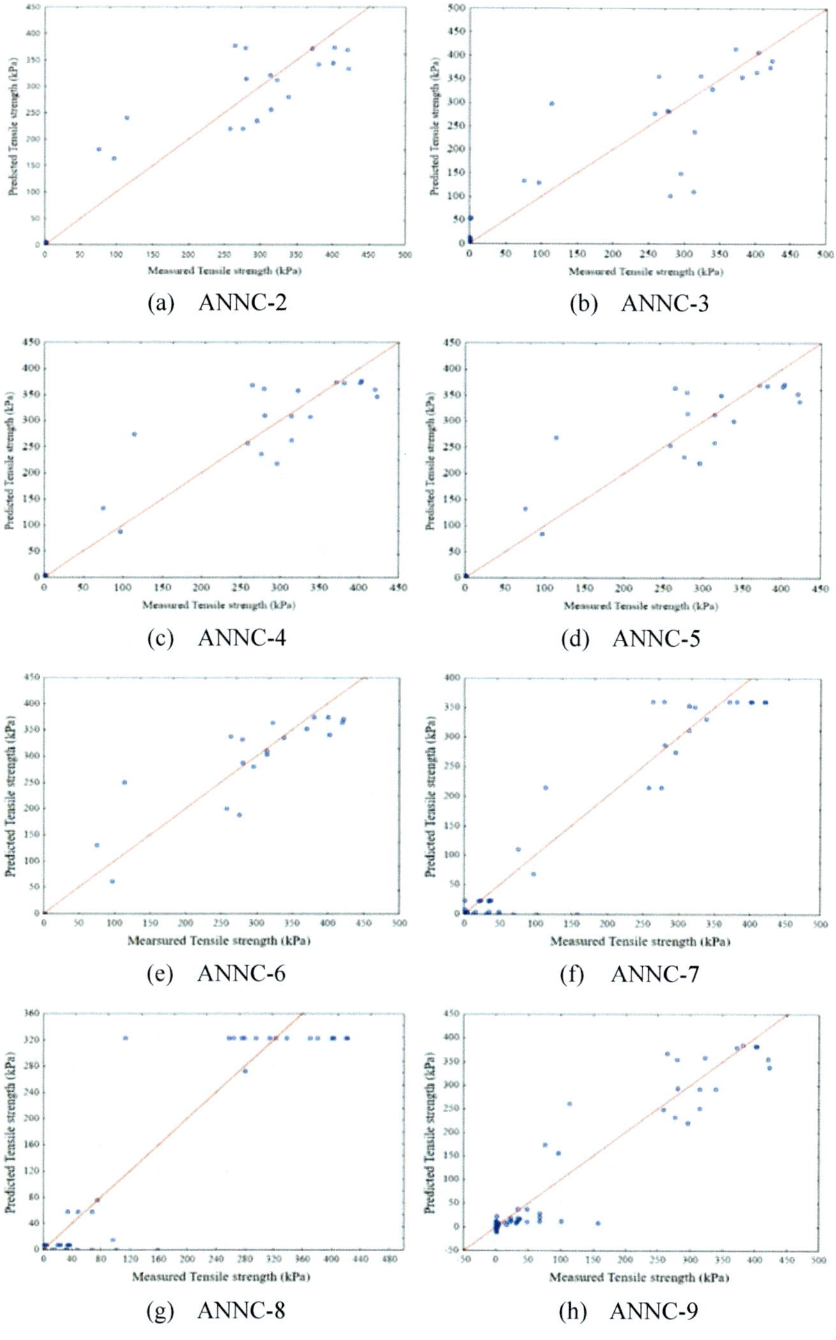

Fig. 2 Measured versus predicted tensile strength of coarse-grained soils using the ANNC-2 to ANN-11 model

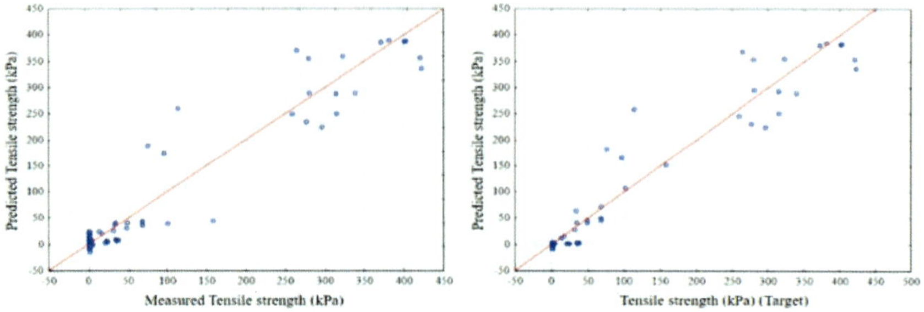

(i) ANNC-10

(j) ANNC-11

Fig. 2 (continued)

on the well-fitting performance of ANN models for coarse-grained and fine-grained soils, sensitivity analyses were programmed to determine the contribution of each parameter on different models as aforementioned.

Table 7 shows the percentage contribution of a parameter obtained from the sensitivity analysis on fine-grained soils. ANNF-1, 2 and 6 indicate that suction plays a dominate role in the tensile strength prediction of fine-grained soil. It is obvious that suction is one of the fundamental properties of fine-grained soils. Therefore, its influence on tensile strength is as expected. As per ANNF-3, in spite of its poor performance with low R^2, it shows that *PI* has more advantage over *FC* in the sensitivity contribution. It is worth mentioning here that both *PI* and *FC* are interrelated as the former parameter originates from the latter component. Similar lines, in both ANNF-4 and ANNF-5 models, PI ranked no. 1. In the former model, *PI* and *FC* are found to have a greater influence on tensile strength than dry density. While in the latter model, the parameter saturation is found to superseding dry density and *FC* on influencing σ_t. When compared the influence between *PI* and ψ on tensile strength prediction, it is clear that suction is prominently influential. From ANNF-7 to ANNF-9, wherein porosity is added as an additional parameter, it showed a relative significance with *PI* andψ. It is intriguing to note from these models that both *PI* and ψ still governs the prediction of tensile strength over porosity. ANNF-9 was developed accounting for all the parameters aforementioned and contribution ranking in the descending order is measured as fines content, suction, saturation, plasticity index, porosity, and dry density (refer to Table 7). In order to affirm the reliability of statistical model, ANNF-9 is repeated five more times with different random starting weights which could control the robustness of the model in relation to its ability to obtain the information about the relative importance of physical factors affecting the tensile strength. The contribution ranking of each parameter in predicting tensile strength of fine-grained soils is summarized in Table 8. The results clearly reveal that the fines content is the most influential parameter (as ranked no. 1) of tensile strength, followed by suction, plasticity index, saturation and then porosity. Whereas, dry density found to be the least important factor influencing the tensile strength of fine-grained soils.

Table 7 Determination of percentage contribution by a parameter from the sensitivity analysis carried out on fine-grained soils

Model name	Contribution (%)
ANNF-1	Suction = 100
ANNF-2	Suction: Saturation = 85.55: 14.45
ANNF-3	Plasticity index: Fines content = 53.61: 46:39
ANNF-4	Plasticity index: Fines content: Dry density = 43.81: 37.59: 18.60
ANNF-5	Plasticity index: Saturation: Dry density: Fines content = 40.82: 39.19: 18.04: 1.96
ANNF-6	Suction: Fines content = 86.19: 13.81
ANNF-7	Plasticity index: Porosity: Saturation: Dry density: Fines content = 43.14: 22.43: 22.13: 7.96: 4.34
ANNF-8	Suction: Plasticity index: Porosity: Dry density: Fines content = 31.91: 29.92: 28.65: 9.39: 0.13
ANNF-9	Fines content: Suction: Saturation: Plasticity index: Porosity: Dry density = 34.24: 26.99: 14.69: 11.53: 11.00: 1.54

Table 8 Results of sensitivity analysis demonstrating the ranking of parameter carried out by ANNF-9 model for fine-grained soils

Trial no	Relative importance of input variable (%)					
	FC	Ψ	PI	S_r	n	ρ_{dry}
1	13.60	5.55	18.70	21.10	18.31	22.75
2	14.69	34.24	26.99	11.53	11.00	1.54
3	20.21	31.88	6.79	24.86	8.42	7.84
4	23.54	25.11	26.97	13.61	9.77	1.00
5	72.09	3.27	16.08	4.38	1.73	2.45
Average	28.82	20.01	19.11	15.10	9.85	7.12
Ranking	1	2	3	4	5	6

Determination of percentage contribution by a parameter from the sensitivity analysis carried out for coarse-grained soils is shown in Table 9. With excellent reliability shown in Fig. 2a–j of the models for coarse-grained soils, further sensitivity analysis was performed. ANNC-2 shows the relative density leads a major role in the tensile strength contribution. Porosity together with relative density shared the same role in ANNC-3, with saturation having the least importance. One of grain size characteristic parameters (i.e. sand content) is added to discuss in ANNC-4 and is indicated as a relative significance second to porosity and relative density, while saturation still played the least important role. When replaced sand content with D_{50} in ANNC-5, it has been noticed that the average particle diameter became the domain parameter for predicting tensile strength with contribution ranking of porosity standing at second and relative density as the least important parameters.

Table 9 Determination of percentage contribution by a parameter from the sensitivity analysis carried out on coarse-grained soils

Model name	Contribution (%)
ANNC-2	Relative density: Saturation = 88.43: 11.57
ANNC-3	Porosity: Relative density: Saturation = 47.28: 46.45: 6.26
ANNC-4	Porosity: Relative density: Sand content: Saturation = 62.56: 23.00: 13.27: 1.17
ANNC-5	D_{50}: Porosity: Saturation: Relative density = 54.20: 45.00: 0.57: 0.23
ANNC-6	Sand content: Porosity: D_{50}: Relative density: Saturation = 25.61: 25.61: 25.61: 16.44: 6.73
ANNC-7	Porosity: Dry density: Saturation = 46.16: 28.52: 25.32
ANNC-8	Porosity: Dry density: D_{50} = 49.61: 45.13: 5.26
ANNC-9	Porosity: Saturation: Dry density: D_{50} = 46.35: 44.57: 5.81: 3.27
ANNC-10	Porosity: Saturation: Sand content: Dry density = 43.16: 24.56: 19.34: 12.93
ANNC-11	Porosity: Saturation: Sand content: Dry density: D_{50} = 43.45: 34.67: 15.23: 4.45: 2.21

When considered simultaneously all of the parameters mentioned above, ANNC-6 is modelled to expose their own sensitivity for predicting tensile strength. It is intriguing to note that parameters relevant to grain size characteristics such as sand content, porosity, and average particle size, manifested considerable significance on the tensile behavior. Amongst all parameters, the contribution of saturation seems to be negligible. Relative density is treated as a synthesis of dry density and porosity. Hence, in the subsequent models (ANNC-7 to ANNC-11), density and porosity were used as a substitute to relative density to rebuild models based on ANNC-2 to ANNC-6. Because the contribution of relative density is released into porosity and dry density, porosity contribution in ANNC-7 to ANNC-11 models enhanced greatly and presents a notable role in tensile strength prediction. Saturation found to promote its own ranking, while the grain size characteristic parameters such as sand content and average particle size faded their fundamental influence as least important parameters. For reproducibility of statistical models, repeated five times of ANNC-6 model with different random starting weights to control the robustness of the model in relation to its ability to obtain the information about the relative importance of the physical factors affecting the tensile strength of soils. The contribution of predicting tensile strength of soil for each parameter is summarized in Table 10, and it distinctly reveals that the sand content is the chief influencing parameter of tensile strength, followed by porosity, average particle diameter and then dry density. Saturation found to be the least important factor influencing the tensile strength of coarse-grained soils.

Table 10 Results of sensitivity analysis demonstrating the ranking of parameter carried out by ANNC-6 model for coarse-grained soils

Trial no	Relative importance of input variable (%)				
	SC	*n*	D_{50}	D_r	S_r
1	28.27	28.27	28.27	27.40	4.26
2	26.72	26.72	26.72	14.29	5.23
3	17.37	17.37	17.37	14.82	1.14
4	26.00	25.99	25.99	15.63	4.74
5	24.31	24.23	22.60	9.56	3.58
Average	24.53	24.52	24.19	16.34	3.79
Ranking	1	2	3	4	5

5 Discussion

Many studies demonstrate that grain size is one of the core properties play a key role in inducing stress–strain and strength behavior to a soil. Vangla and Latha [54] reported that size and morphological characteristics of particles play a vital role in shear and interfacial shear strength of sands. Daphalapurkar et al. [55] used nano-indentation and finite element methods to represent that Young's modulus, hardness, fracture toughness, and stress-strain relations of sand grains depend on the particle size and their distribution characteristics. The sensitivity analysis results presented in Table 10, which accord highest and almost equal ranking to SC and D50 parameters, are in well agreement with the above findings, specifically relevant to the sands, substantiating a fact that tensile strength also depends on percent fraction and particle size distribution characteristics.

A study conducted by Tamrakar et al. [56] on fine-grained soils inferred that with decrease in fraction of fine particles, there is an increase in tensile strength. The variation in the particle size inherently alters the fabric microstructure and thus, the strength of a material such as compressive, shear, tensile and other mechanical properties. ANNF-9 model results (refer to Table 8) discussed herein, which statistically brings out that *FC* is in the foremost position of affecting the tensile strength of fine-grained soils, excellently verify and confirm the above findings.

The findings of other research works noticed to be an excellent match with the current sensitivity analysis results. In line with the above statement, Tamrakar et al. [57] showed an increase in tensile strength and strength ratio with an increase in percentage of fines and decrease in size of finer particles. Lu et al. [49] has also reported that a correlation exists between particle size and tensile strength within the pendular water dominated region and showed that smaller particle sizes exhibit significantly greater tensile strengths.

Many literature works also attribute the development of tensile strength in soils to microstructural hydromechanics and interfacial mechanical interactions. The following studies have contributed on this mechanism and its further understanding.

Munkholm et al. [58] reported the tensile strength of the soil cores was negatively correlated to the macro-porosity of the soil, and was positively correlated to pore continuity and pore organization. Microstructure and microstructural hydromechanics in soils raised the attention of Tang et al. [59]. Several studies, Kim et al. [1], Ajaz and Parry [39] and Narain and Rawat [60] explained that the tensile strength arises from capillary and other pore-scale mechanisms. Lu et al. [49] considered pore size distribution, fully described by two parameters: the inverse value of the air-entry pressure and the pore size spectrum parameter, as a fundamental control for tensile strength.

In the context of above contrasting discussions, it is difficult to pinpoint the exact contributing factors on the development of tensile strength in soils. At this juncture, the sensitivity analysis, which has been carried out categorically taking into account for one by one parameter, aids in simplifying the parametric effect on the tensile strength.

As such, the model results of ANNF-0, 1, & 3 and ANNC-0 & 1 ratify that the consideration given exclusively to internal or external parameters has no benefit in tensile strength prediction. In similar lines, ANNF-9 (Table 8) and ANNC-6 (Table 10) models grant highest ranking to grain size characteristics, in the presence of both internal and external parameters. For fine-grained soils, in spite of fines content, plasticity index, which in turn conveys the grain size information, play the most important role in ANNF-3, ANNF-4, ANNF-5, and ANNF-7 models. These findings corroborate well with the results of Kim et al. [1] and Fang and Chen [61], who highlighted that PI increases so does the tensile strength. Suction, induced by pore-scale mechanism from microstructural hydromechanics and interfacial mechanical interactions, apparently governs tensile strength as indicated by ANNF-1, ANNF-2, ANNF-6, and ANNF-8 models. These observations match very well with similar conclusions derived by Zeh and Witt [14, 15], Trabelsi et al. [21] and Bardanis and Grifiza [38]. With regard to porosity, the studies in similar lines are by Zeh and Witt [15], Trabelsi et al. [21], which played relatively an important role in ANNF-7 to ANNF-9 models. For coarse-grained soils, porosity and relative density as macro-parameters affected tensile strength greatly, as evident from ANNC-2 to ANNC-4 and ANNC-7 to ANNC-11. When bunched all factors, ANNC-5 and ANNC-6 present inherent function of tensile strength with average particle diameter and sand content. This observation well validates the study by Yin and Vanapalli [62], who used SWCC as a tool to propose a model involving average particle size and coefficient of uniformity to predict tensile strength of unsaturated cohesionless soils.

Based on the findings discussed herein, an attempt is also made to link the micro-interaction mechanism of grain size parameters with relevant studies. As such, the results of ANNF-9 demonstrate that fines content, the contribution of which is superior to over its counterpart parameters (refer to Table 8), plays a prominent role in the evolution of tensile strength of fine-grained soils. This observation excellently corroborates with discussion made by Nearing et al. [2], who have reported that clay content and surface properties are dominant parameters in controlling the tensile strength of loose saturated soils, and Barzegar et al. [3], who have also concluded that the tensile strength increases with an increase in the clay fraction. This is because clay content is closer to the explanation of microstructural hydromechanics and interfacial mechanical interactions than fines content.

Based on elaborate discussions and demonstrations from several ANN models presented in the paper, it is prudent to confirm that grain size parameters are the key factors in the evolution of tensile strength in the case of both fine-grained and coarse-grained soils. When interpretations of various ANN model results integrated with literature revelations, the role of microstructural hydromechanics and interfacial mechanical interactions in the evolution of tensile strength cannot also be undermined.

6 Conclusions

In the present study, 22 ANN models including 10 for fine-grained soils and 12 for coarse-grained soils were developed with an aim to comprehend the exact effect of a specific parameter on the tensile behavior of these soils. The analyses, evidently, manifest that it is essential to incorporate both internal and external parameters for accurate prediction of the tensile strength of soils. Series of sensitivity analyses affirm that ANN is one of the best tools for recognizing the prominently influential parameter when there is a more than one input parameter is involved in the modeling. It is intriguing to note that fines content, suction and plasticity index in the case of fine-grained soils, and sand content, porosity and average particle diameter in the case of coarse-grained soils are the profoundly influential parameters of the tensile strength. Overall, the results portray that for reliable prediction of tensile strength, parameters relevant to grain size properties are essential to be incorporated in the analysis. This study is first of its kind in exploring the relative importance of parameters influencing tensile strength of coarse- and fine-grained soils. Collection of the relevant database from the literature is only the constraint in the study, which can be overcome by resorting to series of systematic experimental studies under controlled parametric conditions, in the future, for evaluating the tensile strength of fine-grained and coarse-grained soils. This will also help in generating the database that is more methodical with less non-uniformity in material, input parameters or boundary conditions.

APPENDIX-I: MATLAB Script Written by the Author

```
% Solve an Input-Output Fitting problem with a Neural
Network
% Script generated by NFTOOL
% Created Thu Mar 30 02:01:36 IST 2017
%
% This script assumes these variables are defined:
%
%finetwo - input data.
%finetwooutput - target data.
inputs = finetwo';
targets = finetwooutput';
% Create a Fitting Network
hiddenLayerSize = 3;
net = fitnet(hiddenLayerSize);
% Choose Input and Output Pre/Post-Processing Functions
% For a list of all processing functions type: help
nnprocess
net.inputs{1}.processFcns = {'removeconstan-
trows','mapminmax'};
net.outputs{2}.processFcns = {'removeconstan-
trows','mapminmax'};
% Setup Division of Data for Training, Validation, Test-
ing
% For a list of all data division functions type: help
nndivide
net.divideFcn = 'dividerand';  % Divide data randomly
net.divideMode = 'sample';  % Divide up every sample
net.divideParam.trainRatio = 50/100;
net.divideParam.valRatio = 25/100;
net.divideParam.testRatio = 25/100;
% For help on training function 'trainlm' type: help
trainlm
```

```
% For a list of all training functions type: help nntrain
net.trainFcn = 'trainlm';  %Levenberg-Marquardt
% Choose a Performance Function
% For a list of all performance functions type: help
nnperformance
net.performFcn = 'mse';  % Mean squared error
% Choose Plot Functions
% For a list of all plot functions type: help nnplot
net.plotFcns = {'plotper-
form','plottrainstate','ploterrhist', ...
  'plotregression', 'plotfit'};
% Train the Network
[net,tr] = train(net,inputs,targets);
% Test the Network
outputs = net(inputs);
errors = gsubtract(targets,outputs);
performance = perform(net,targets,outputs)
% Recalculate Training, Validation and Test Performance
trainTargets = targets .* tr.trainMask{1};
valTargets = targets  .* tr.valMask{1};
testTargets = targets  .* tr.testMask{1};
trainPerformance = perform(net,trainTargets,outputs)
valPerformance = perform(net,valTargets,outputs)
testPerformance = perform(net,testTargets,outputs)
% View the Network
view(net)
% Plots
% Uncomment these lines to enable various plots.
%figure, plotperform(tr)
%figure, plottrainstate(tr)
%figure, plotregression(targets,outputs)
```

References

1. Kim TH, Kim TH, Kang GC, Ge L (2012) Factors influencing crack-induced tensile strength of compacted soil. J Mater Civ Eng 24(3):315–320
2. Nearing MA, Parker SC, Bradford JM, Elliot WJ (1991) Tensile strength of thirty-three saturated repacked soils. Soil Sci Soc Am J 55(6):1546–1551. https://doi.org/10.2136/sssaj1991.03615995005500060008x
3. Barzegar AR, Oades JM, Rengasamy P, Murray RS (1995) Tensile strength of dry, remoulded soils as affected by properties of the clay fraction. Geoderma 65(1–2):93–108. https://doi.org/10.1016/0016-7061(94)00028
4. Wang JJ, Zhu JG, Chiu CF, Zhang H (2007) Experimental study on fracture toughness and tensile strength of a clay. Eng Geol 94(1–2):65–75. https://doi.org/10.1016/j.enggeo.2007.06.005

5. Fang HY, Fernandez J (1981) Determination of tensile strength of soils by unconfined-penetration test. In Laboratory shear strength of soil, ASTM International
6. Uday KV, Jayanthi PN, Singh DN (2014) A generalized relationship for determination of tensile strength of fine-grained soils from shrinkage characteristics. Drying Technol 32(7):869–876. https://doi.org/10.1080/07373937.2013.873804
7. Vanicek I (2013) The importance of tensile strength in geotechnical engineering. Acta Geotech Slov 1:5–17
8. Rosenqvist IT (1959) Physico-chemical properties of soils: soil-water systems. J Soil Mech Found Div 85(2):31–54
9. Bishop AW, Garga VK (1969) Drained tension tests on London clay. Geotechnique 19(2):309–313
10. Bishop AW (1966) The strength of soils as engineering materials. Geotechnique 16:202–232
11. Parry RHG, Nadarajah V (1974) Anisotropy in a natural soft clayey silt. Eng Geol 8(3):287–309. https://doi.org/10.1016/0013-7952(74)90003-9
12. Causarano H (1993) Factors affecting the tensile strength of soil aggregates. Soil Tillage Res 28(1):15–25. https://doi.org/10.1016/0167-1987(93)90052-Q
13. Tang GX, Graham J (2000) A method for testing tensile strength in unsaturated soils. Geotech Test J 23(3):377–382
14. Zeh RM, Witt KJ (2005) Suction-controlled Tensile Strength of Compacted Clays. In: Proceedings of the International Conference on Soil Mechanics and Geotechnical Engineering, vol 4, pp 2347–2352
15. Zeh RM, Witt KJ (2007) The tensile strength of compacted clays as affected by suction and soil structure. Exp Unsaturated Soil Mech, Springer Proc Phys 112(3):219–226. https://doi.org/10.1007/3-540-69873-6_21
16. Win SS (2006).Tensile strength of compacted soils subject to wetting and drying. Master Thesis, University of New South Wales,Sydney, Australia
17. Dexter AR, Chant KY (1991) Soil mechanical properties as influenced by exchangeable cations. J Soil Sci 42(2):219–226. https://doi.org/10.1111/j.1365-2389.1991.tb00403.x
18. Greene RS, Eggleton RA, Rengasamy P (2002) Relationships between clay mineralogy and the hard setting properties of soils in the Carnarvon horticultural district of Western Australia. Appl Clay Sci 20(4–5):211–223. https://doi.org/10.1016/S0169-1317(01)00073-4
19. Oliveira PV, Correia AAS, Teles JMNPC, Custódio DG (2016) Effect of fibre type on the compressive and tensile strength of a soft soil chemically stabilised. Geosynth Int 23(3):171–182. https://doi.org/10.1680/jgein.15.00040
20. Consoli NC, de Moraes RR, Festugato L (2011) Split tensile strength of monofilament polypropylene fiber-reinforced cemented sandy soils. Geosynth Int 18(2):57–62. https://doi.org/10.1680/gein.2011.18.2.57
21. Trabelsi H, Jamei M, Zenzri H, Olivella S (2012) Crack patterns in clayey soils: experiments and modelling. Int J Numer Anal Meth Geomech 36(11):1410–1433. https://doi.org/10.1002/nag.1060
22. Tej PR, Singh DN (2012) Estimation of tensile strength of soils from penetration resistance. Int J Geomech 13(5):496–501
23. Barzegar AR, Murray RS, Churchman GJ, Rengasamy P (1994) The strength of remolded soils as affected by exchangeable cations and dispersible clay. Soil Res 32(2):185–199. https://doi.org/10.1071/SR9940185
24. Ibarra SY, McKyes E, Broughton RS (2005) Measurement of tensile strength of unsaturated sandy loam soil. Soil Tillage Res 81(1):15–23. https://doi.org/10.1016/j.still.2004.04.002
25. Venkataramana K, Rao BH, Singh DN (2009) A critical review of the methodologies employed for determination of tensile strength of fine-grained soils. J Test Eval 37(2):115–121
26. Shahin MA, Jaksa MB, Maier HR (2001) Artificial neural network applications in geotechnical engineering. Aust Geomech 36(1):49–62
27. Choobbasti AJ, Farrokhzad F, Barari A (2009) Prediction of slope stability using artificial neural network (case study: Noabad, Mazandaran, Iran). Arab J Geosci 2(4):311–319. https://doi.org/10.1007/s12517-009-0035-3

28. Yilmaz I, Yuksek AG (2008) An example of artificial neural network (ANN) application for indirect estimation of rock parameters. Rock Mech Rock Eng 41(5):781–795. https://doi.org/10.1007/s00603-007-0138-7
29. Abdel-Rahman AH (2008) Predicting compaction of cohesionless soils using ANN. Proc Inst Civ Eng-Ground Improv 161(1):3–8. https://doi.org/10.1680/grim.2008.161.1.3
30. Kumar S, Basudhar PK (2018) A neural network model for slope stability computations. Géotechnique Lett 8(2):149–154. https://doi.org/10.1680/jgele.18.00022
31. Soleimanbeigi A, Hataf N (2006) Prediction of settlement of shallow foundations on reinforced soils using neural networks. Geosynth Int 13(4):161–170. https://doi.org/10.1680/gein.2006.13.4.161
32. Tang CS, Wang DY, Cui YJ, Shi B, Li J (2016) Tensile strength of fiber-reinforced soil. J Mater Civ Eng 28(7):04016031
33. Li J, Tang C, Wang D, Pei X, Shi B (2014) Effect of discrete fibre reinforcement on soil tensile strength. J Rock Mech Geotech Eng 6(2):133–137. https://doi.org/10.1016/j.jrmge.2014.01.003
34. Zhang BY, Li QM, Yuan HN, Sun X (2014) Tensile fracture characteristics of compacted soils under uniaxial tension. J Mater Civ Eng 27(10):04014274
35. Stirling RA, Hughes P, Davie CT, Glendinning S (2015) Tensile behaviour of unsaturated compacted clay soils-A direct assessment method. Appl Clay Sci 112:123–133. https://doi.org/10.1016/j.clay.2015.04.011
36. Beckett CTS, Smith JC, Ciancio D, Augarde CE (2015) Tensile strengths of flocculated compacted unsaturated soils. Géotechnique Lett 5(4):254–260. https://doi.org/10.1680/jgele.15.00087
37. Ramanathan B, Raman V (1974) Split tensile strength of cohesive soils. Soils Found 14(1):71–76. https://doi.org/10.3208/sandf1972.14.71
38. Bardanis M, Grifiza S (2012) Soil-water characteristic curves and evolution of the tensile and unconfined compression strength of drying slurried soils. In: Proceedings of the unsaturated soils: research and applications. Springer, Berlin, Heidelberg, pp 249–256. https://doi.org/10.1007/978-3-642-31116-1_34
39. Ajaz A, Parry RH (1975) Stress–strain behaviour of two compacted clays in tension and compression. Geotechnique 25(3):495–512. https://doi.org/10.1680/geot.1975.25.3.495
40. Zeh RM, Witt KJ (2006) Tensile strength of compacted clays as part of the crack prediction of clay liner in landfills. In: 5th ICEG environmental geotechnics: opportunities, challenges and responsibilities for environmental geotechnics: proceedings of the issmge's fifth international congress organized by the geoenvironmental research centre. Cardiff University, Cardiff City Hall, pp 816–823
41. Nahlawi H, Chakrabarti S, Kodikara J (2004) A direct tensile strength testing method for unsaturated geomaterials. Geotech Test J 27(4):356–361. https://doi.org/10.1520/GTJ11767
42. Lakshmikantha MR, Prat PC, Ledesma A (2012) Experimental evidence of size effect in soil cracking. Can Geotech J 49(3):264–284. https://doi.org/10.1139/t11-102
43. Kim TH, Hwang C (2003) Modeling of tensile strength on moist granular earth material at low water content. Eng Geol 69(3–4):233–244. https://doi.org/10.1016/S0013-7952(02)00284-3
44. Lu N, Wu B, Tan CP (2007) Tensile strength characteristics of unsaturated sands. J Geotech Geoenvironmental Eng 133(2):144–154
45. Araki H, Koseki J, Sato T (2016) Tensile strength of compacted rammed earth materials. Soils Found 6(2):189–204. https://doi.org/10.1016/j.sandf.2016.02.003
46. Farrell DA, Greacen EL, Larson WE (1967) The effect of water content on axial strain in a loam soil under tension and compression. Soil Sci Soc Am J 31(4):445–450. https://doi.org/10.2136/sssaj1967.03615995003100040011x
47. Jindal P, Sharma J, Bashir R (2016) Effect of pore-water surface tension on tensile strength of unsaturated sand. Indian Geotech J 46(3):276–290. https://doi.org/10.1007/s40098-016-0184-8
48. Kim T, Sture S, Yun J (2004) Investigation of moisture effect on tensile strength in granular soil. In: Proceedings of the 9th biennial ASCE aerospace division international conference on engineering, construction, and operations in challenging environments, pp 57–64

49. Lu N, Kim TH, Sture S, Likos WJ (2009) Tensile strength of unsaturated sand. J Eng Mech 135(12):1410–1419
50. Goh AT (1995) Back-propagation neural networks for modeling complex systems. Artif Intell Eng 9(3):143–151. https://doi.org/10.1016/0954-1810(94)00011-S
51. Twomey M, Smith AE (1997) Validation and verification. Fundamentals and applications, ASCE, New York, Artificial neural networks for civil engineers, pp 44–64
52. Demuth H, Beale M, Hagan M (2006) Neural network toolbox user's guide. The Math Works Inc, Natick, MA 01760–2098
53. Garson GD (1991) Interpreting neural-network connection weights. AI Expert 6(4):46–51
54. Vangla P, Latha GM (2015) Influence of particle size on the friction and interfacial shear strength of sands of similar morphology. Int J Geosynth Ground Eng 1(1):6. https://doi.org/10.1007/s40891-014-0008-9
55. Daphalapurkar NP, Wang F, Fu B, Lu H, Komanduri R (2011) Determination of mechanical properties of sand grains by nanoindentation. Exp Mech 51(5):719–728. https://doi.org/10.1007/s11340-010-9373-z
56. Tamrakar SB, Toyosawa Y, Mitach T, Itoh K (2005) Tensile strength of compacted and saturated soils using newly developed tensile strength measuring apparatus. Soils Found 45(6):103–110
57. Tamrakar SB, Mitachi T, Toyosawa Y (2007) Measurement of soil tensile strength and factors affecting its measurements. Soils Found 47(5):911–918. https://doi.org/10.3208/sandf.47.911
58. Munkholm LJ, Schjønning P, Kay BD (2002) Tensile strength of soil cores in relation to aggregate strength, soil fragmentation and pore characteristics. Soil Tillage Res 64(1–2):125–135. https://doi.org/10.1016/S0167-1987(01)00250-1
59. Tang CS, Pei XJ, Wang DY, Shi B, Li J (2014) Tensile strength of compacted clayey soil. J Geotech Geoenvironmental Eng 141(4):04014122
60. Narain J, Rawat PC (1970) Tensile strength of compacted soils. J Soil Mech Found Div 96:2185–2190
61. Fang HY, Chen WF (1971) New method for determination of tensile strength of soils. Highway Research Board, National Research Council, Washington, DC, pp 62–68
62. Yin P, Vanapalli SK (2018) Corrigendum: model for predicting tensile strength of unsaturated cohesionless soils. Can Geotech J 56(1):153–153. https://doi.org/10.1139/cgj-2018-0785

Performance of Retaining Walls Backfilled with Blend of Sand and Building Derived Materials: A Laboratory Scale Study

Anasua GuhaRay and M. Jayatheja

Abstract Disposal of the humungous quantity of generated construction and demolition debris is a serious problem in the present world. Although sands are considered as the most suitable backfill for retaining walls due to its high permeability, nowadays the scarcity of this natural material has raised serious environmental concerns. The present study aims at the sustainable and eco-friendly use of building derived materials (BDM) as a partial replacement of backfill material for earth retaining structures. An attempt is made to study the feasibility of utilizing BDM and its effect on at-rest, active and passive earth pressures on rotational mode of failure. Small scale laboratory model tests are conducted on a cantilever rigid retaining wall with different soil-BDM blends as backfill. The experimental results indicate that the earth pressures are not increased significantly by the inclusion of BDM to soil. For passive conditions, since the wall is made to rotate about its bottom, greater values of earth pressure is observed near the upper part of the wall and it decreases non-linearly with depth. Numerical simulations are conducted in line with the experiments to validate the observations. Since the numerical results suggest a good agreement with that of experimental results, it may be said that the finite element model provides enough support to validate the experimental outcomes. This model can be further used for research to evaluate the behavior of backfill soil with different BDM content. The present research is a step towards the sustainable and environment-friendly solution for recycling building derived waste material and has a significant practical impact on low cost construction of earth retaining structures.

Keywords Building derived materials · Lateral earth pressures · Laboratory scale retaining wall tests · Hyperbolic model · Finite element method

A. GuhaRay (✉) · M. Jayatheja
BITS Pilani Hyderabad Campus, Secunderabad 500078, Telangana, India
e-mail: guharay@hyderabad.bits-pilani.ac.in

A. Garg et al. (eds.), *Proceedings of the 1st Indo-China Research Series in Geotechnical and Geoenvironmental Engineering*, Lecture Notes in Civil Engineering 123,
https://doi.org/10.1007/978-981-33-4324-5_5

1 Introduction

The basic principles of 'Sustainable Development' were formulated originally by Hans Carl von Carlowitz, a forester in Germany, who defined sustainability as "sparing use of trees to give the forest a chance to regenerate and sustain itself" [16, 17]. In the modern age of rapid urbanization and infrastructural development, large quantities of waste mate-rials are generated at a global level, annually. Environmental Protection Agency (EPA) defines "construction and demolition debris as the waste material that is produced in the process of construction, renovation or demolition of structures" [27, 28]. Typically, the waste materials derived from the building demolition or renovation includes materials such as crushed concrete, asphalt, wood, crushed brick, broken metals and roofing shingles. Moreover, the demolition of concrete structures and disposal of the associated wastes is an environmental problem. Rejected construction materials and building derived materials (BDM) constitutes the single largest portion of these wastes, amounting to 3 billion tonnes per annum worldwide [1]. In USA, out of total generated MSW, 52.1% of solid waste materials are landfilled and 25.1 are recycled [29]. Thus, it is imperative to adopt innovative techniques to utilize rejected concrete wherever applicable, instead of producing fresh concrete.

According to 2019 statistics, Asia–Pacific region is the largest producer of concrete and cement. China and India produced nearly 2500 Million metric tons of CDW in the year 2019 [7]. Moreover, it is estimated that one ton of CO_2 is produced for manufacturing one ton of Portland cement [27]. This accounts for 4–8% of the global annual CO_2 production, originated from manufacturing process of clinker material.

Meyer [17], Limbachiya et al. [13] there are several possible ways of addressing the above issues. Some of the techniques in vogue are: (i) Increased use of supplementary cementitious mate-rials to reduce the demand for Portland cement and subsequently reducing CO_2 emissions; (ii) Increased reliance on recycled materials as aggregate in concrete. In the webinar 'Sustainable Geotechnical Applications' conducted by ASCE, Melton [15] and Edil and Bosscher [5] suggested that the strength characteristics of soil can be substantially improved by mixing recycled concrete into them. Lin et al. [14] studied experimentally the effect of mixing recycled concrete aggregate on strength and durability of flyash based geopolymer concrete. These studies concluded that the durability due to addition of recycled concrete is similar to normal limestone geopolymer concrete. Silva et al. [21] carried out a meta-analysis from published literature on the feasibility of using aggregates from recycled construction and demolition wastes on the performance of masonry mortars. They concluded that, an increase in the amount of recycled aggregate reduced the performance of masonry mortars in some cases, while it showed comparable performance in rendering applications to that of conventional mortars. Hasan et al. [8] studied the effect of ground granulated blast furnace slag and recycled construction waste on stabilization of bentonite clay. The authors suggested that a combination of 5% slag and 20% construction waste can be considered as optimum mix. In India, very limited measures have been adopted to recycle and reuse the construction and demolition wastes [12, 23, 22]. Rao et al. [19] investigated the influence of different amounts of recycled coarse aggregates on the properties of recycled aggregate concrete. They developed the relationships among compressive strength, tensile strengths and elastic modulus, which were verified with the available models for normal and recycled aggregate concrete.

However, some practical challenges are associated with the process of recycling any waste material. Procurement of the material, processing, transportation, and the relevant financial investments have to be considered. But as mentioned above, concrete debris is available in abundance, which makes them a suitable option for recycling and reuse. Existing re-search shows the use of such debris to produce new concrete, thereby conserving natural resources of limestone and reducing the disposal of these wastes into landfills [17]. Construction and demolition waste generation in India accounts up to 180 million tons annually and these figures are likely to increase up to 200 million tons by 2025 [10]. It has become essential to study its generation and handling to capture an accurate estimate and establish sustainable methodologies to recycle these wastes. The construction materials account for nearly half of the average construction costs [18].

"Over the last five years, India's first and only recycling plant for construction and demolition (C&D) waste has saved the already-polluted Yamuna and the overflowing landfills of Delhi from 15.4 lakh tonnes of debris. A circular from the Ministry of Urban Development dated June 28, 2012, directed the states to set-up such facilities in all cities with a population of over 10 lakhs. But, till now the existing facility at Burari is the only one."—The Hindu [26]. In pursuant to the Working Committee Report on MSW Management 2010 and Draft SWM Rules, 2015 of MoEF&CC, C&D Waste Management Rules 2016 have been issued by MoEF&CC vide notification no. G.S.R. 317(E) dated 29th March 2016. Ministry of Urban Development (MoUD) circular dated June 28, 2012 as well as "Swatch Bharat Mission" (BMTPC 2017) directs reuse and recycling of CDW. CPWD and NBCC have also recommended use of recycled CDW products. BIS vide its amendment of 2016 to IS 383 permits use of CDW products in plain and reinforced concretes. Building Materials and Technology Promotion Council (BMTPC), Government of India prepared guidelines suggesting incentivization for use of recycled CDW products up to a minimum of 20% of conventional building materials and in-situ processing of CDW.

India being the second largest populated country in the world, the construction industry in India is one of the major economic sectors after agriculture. It is essential to assess the quantity of construction and demolition waste generated annually. Apart from demolition of infrastructure, occurrence of natural disasters like earthquakes also generate huge demolition debris. Approximately 59% of Indian subcontinent has a potential risk from moderate to severe earthquakes. Major earthquakes in the last 30 years such as Khillari (1993), Jabalpur (1997), Chamoli (1999), Bhuj (2001) and Nepal (2015) have resulted in more than 25,000 deaths and extensive damage to infrastructure (NDMA 2010). These humungous volumes of building debris that are generated due to collapse of infrastructures during earthquakes need to be disposed before commencing any renovation works. Hence, novel techniques need to be proposed and adopted to recycle and reuse these solid wastes.

Sustainability is a very important concept in the present world. The generation and handling issues of construction and demolition waste are in focus nowadays to achieve sustainability. Recycling and reuse of these materials serve bi-fold benefit of aiding the environment by avoiding waste disposal and at the same time, reducing the costs associated with the disposal process. The production and emission of greenhouse gas and other pollutants are minimized, resulting in reduced need to extract and transport raw materials to long distances. At the same time, it reduces the need for new landfills and the costs involved in it. One possible use of this material could be in ground improvement, as suggested by Melton [15] and Edil and Bosscher [5]. Another potential use of demolition waste material is for backfilling of various highway embankments. Presently, geosynthetics reinforced local soils are widely used backfill material. One sustainable and economic solution can be partial or full replacement of backfill soil with BDM. However, a significant amount of research encompassing laboratory investigations and numerical studies is needed before the wide use of waste materials in highway embankments. The emerging interest in utilizing waste materials in geotechnical engineering applications has opened the possibility of constructing retaining structures with unconventional backfills.

Till now, there has been limited significant study taken up towards the use of the building derived material (BDM) along with soil as a backfill material. This study primarily aims to assess the performance of locally available soil backfills, partially replaced with building derived materials, for retaining structures. In the present study, the performance of the BDM-backfill material is investigated. Detailed material and geotechnical characterization are carried out on each of these materials. The strength and volumetric properties of individual material, as well as their mixture in different proportions, are investigated experimentally. A finite element analysis on commercially available software is also carried out based on the experimental results to predict the behavior of BDM-backfill material. The most appropriate mix proportions are determined and reported which will have a significant practical impact on low cost construction of earth retaining structures. The present research is a step towards the sustainable and environment-friendly solution for recycling building derived waste material.

2 Materials

2.1 Sand

The locally available river sand (S), used in the present study, is collected from the East Godavari river basin near Telangana state of India. The grain size analysis of sand shows that it can be classified as a poorly graded sand (SP) according to Unified Soil Classification System (USCS). A medium dense condition is simulated throughout the experimental procedure by maintaining a constant relative density of 44%. This is achieved by rainfall pouring technique from a height of 2 m. A maximum dry unit weight of 18.43 kN/m^3 is obtained by vibrating the sand in a vibratory table. The minimum dry unit weight is 17.24 kN/m^3 obtained by pouring in its loosest state. From the grain size distribution curve (Fig. 2), it is observed that the gradation parameters (Uniformity Coefficient, C_u and Curvature Coefficient C_c) of sand are 2.5 and 0.9 respectively, indicating a poorly graded soil. The sand particles, as observed through stereomicroscopic images, are mostly angular in shape, with smaller percentages of rounded and sub-rounded particles.

2.2 *Building Derived Materials (BDM)*

The BDM (Fig. 1) is composed of waste collected from two local construction and demolition building sites near Kapra Mandal in Telangana, India. The collected CDW is a conglomeration of concrete, brick, glass, wood, and plastic. According to Bhattacharya et al. (2013), CDW consists of 65% concrete, 25% brick, 5% wood and 5% miscellaneous materials like metals and plastic. The concrete and brick component of this CDW is separated manually and used as BDM for this study. A mechanically operated jaw crusher is used to crush the collected BDM to sizes less than 10 mm. ASTM guidelines are followed to maintain consistency during the process of sampling and separation.

Sieve analysis of BDM is carried out according to IS: 2720 (Part 4)—1985. The BDM can be classified as poorly graded gravel (GP) according to Unified Soil Classification System (Fig. 2). The consistency limit tests carried out on finer sized particles of BDM showed that BDM is non-plastic in nature.

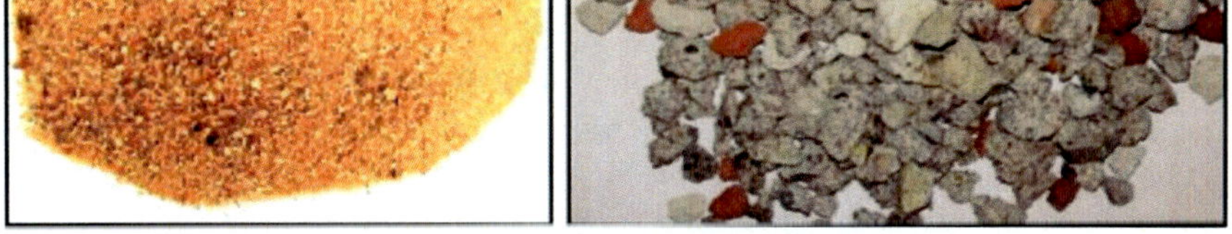

Fig. 1 Sand and building derived materials

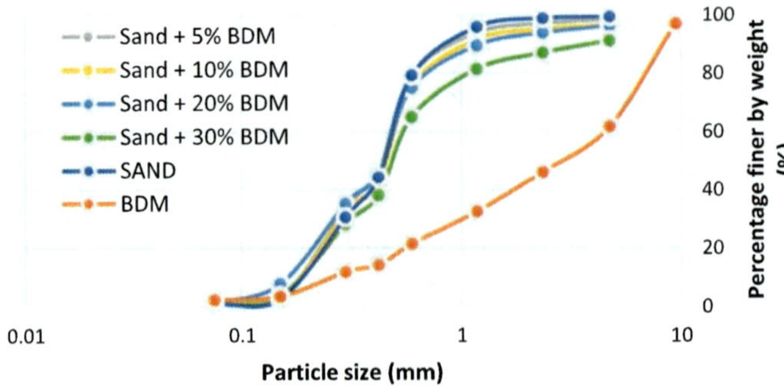

Fig. 2 Particle size distribution of sand blended with different percentages of BDM

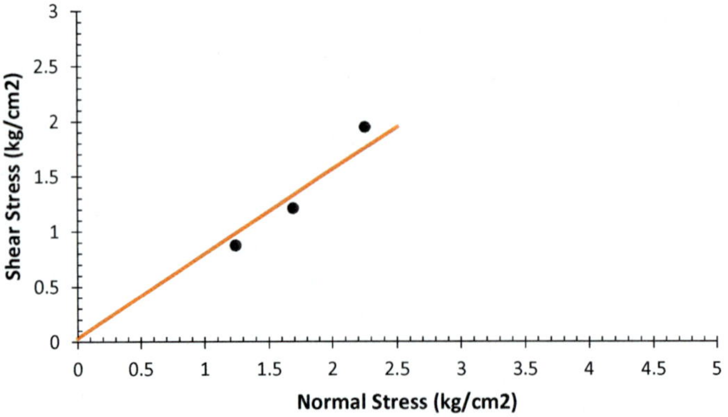

Fig. 3 Normal versus shear stress of building derived material (BDM)

A large direct shear test box with cross-sectional area of 300 mm × 300 mm and depth of 300 mm in is used to determine angle of internal friction (φ) for BDM according to ASTM D3080 (2011) specifications. The BDM samples are sheared at 1.25 kg/cm^2, 1.75 kg/cm^2 and 2.25 kg/cm^2 at a constant strain rate of 0.028 mm/min. The φ of BDM is reported as 46°, as indicated in Fig. 3.

2.3 *Sand-BDM Blend*

Sand (S) is blended with different percentages of BDM (5–30% by weight). The grain size analysis is carried out for the different blends and is presented in Fig. 2. It is observed that the gradation parameters, Cu and Cc of sand, which is initially classified as poorly graded, improved with addition of BDM.

The variation of maximum dry unit weight (MDU) and optimum moisture content (OMC) of sand blended with different percentages of BDM is shown in Fig. 4. It is

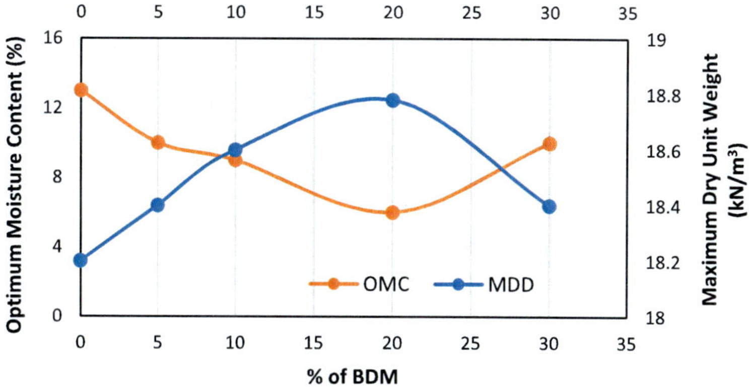

Fig. 4 Variation of MDU and OMC with BDM

observed that as the quantity of BDM is increased in soil, MDU also increases till it reaches its optimum at 20% and reverse trend is noted after optimum. Optimum values of MDU and OMC of sand are obtained at 19.2 kN/m^3 and 8.75% respectively on blending of 20% of BDM.

3 Lateral Earth Pressures from Laboratory Scale Model Retaining Wall Test Set-Up

According to Terzaghi [24], satisfactory values of earth pressures according to classical theories of Coulomb and Rankine can be obtained when shear strength of soil is fully mobilized. Terzaghi [25] also stated out that a retaining wall can fail either by translational or rotational movement of the wall about its top or bottom. The active earth pressure follows a hydrostatic distribution when a retaining wall rotates about its base. However, it is observed that the distribution of earth pressure is nonlinear for other types of movements, which is in contrast to Coulomb or Rankine's theory. Also, the point of application varies for different cases [9, 20]. However, according to Fang and Ishibashi [6], mobilization of active state of rigid retaining walls is independent of the type of wall movement.

3.1 Experimental Set-Up

The retaining wall is built in a steel tank of length 1.22 m, breadth 0.92 m and height 0.92 m (Fig. 5). Three sides and bottom base of the tank are made with mild steel material of 0.012 m thick and one side is fitted with 0.025 m thick acrylic sheets, fitted in position with mild steel bar elements with fasteners. A mild steel plate of

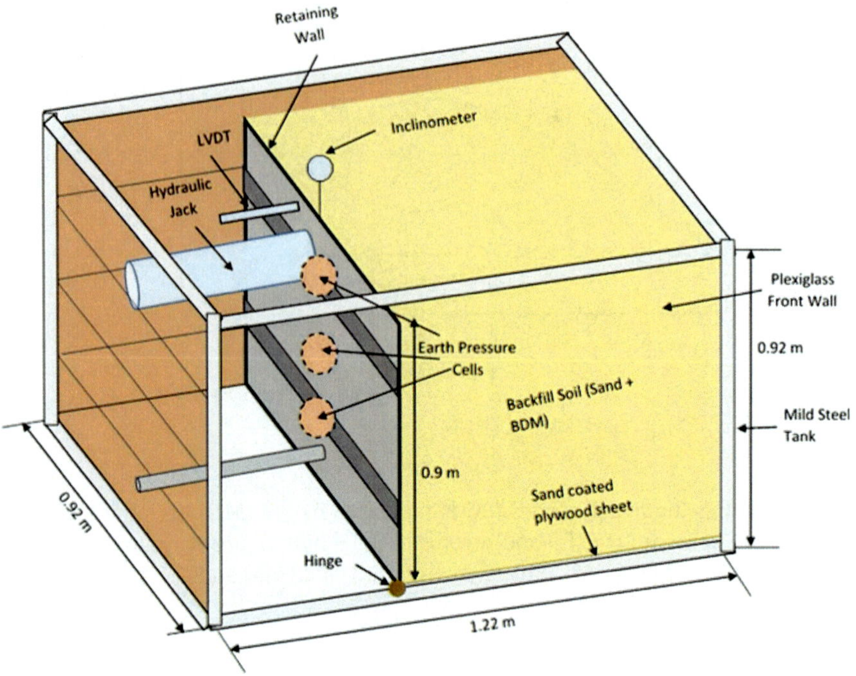

Fig. 5 Schematic diagram of laboratory scale model retaining wall

0.012 m thickness and height 0.9 m placed in the tank over a hinge, acts as the retaining wall. The width of the backfill is maintained as 0.65 m. The soil is confined between these two rigid steel plates that represent the front retaining wall and the back rigid wall, respectively. The various modes of movements of the front retaining wall are simulated by operating the two control units which are mounted on the upper and lower part of the front wall. A hinge is attached to front side at the bottom of the wall through rotational joint which provides a rotation of 20° on two sides. This hinge is connected to a crank with a handle which is threaded and fixed to front side of tank. To capture responses at different cases of rotational failure, a hydraulic jack of 150 kN load capacity is arranged at 0.75 m height. The shaft length is arranged at mean position and active and passive states are achieved when the shaft is pushed towards or away from backfill respectively. To ensure uniformity in load distribution, small plate is mounted on front portion of the retaining wall in contact with hydraulic jack. To capture the variation of magnitude and distribution of lateral pressure on the retaining wall, three earth pressure cells (EPC) of diameter 0.2 m and thickness 0.007 m were fixed at different depths and along the center line of the back of the wall.

The different combinations of backfill material used in the present study are sand blended with 5, 10, 20 and 30% of BDM. The dry pluviation method is adopted to place the backfill from a height of 2 m, thereby maintaining a constant relative

Fig. 6 Movement of backfill soil as observed from front

density of 44%. The backfill is then allowed left undisturbed for 48 h to achieve a relatively uniform compaction. In addition to the measurements of lateral pressure at three locations, the back-fill movement from the top is visually observed to monitor the boundary effects. Figure 6 shows the settlement of the uniform settlement of the back-fill in the z-direction under active earth pressure conditions. This indicates that the boundary effects in the z- dimension as well as the side wall frictions are negligible. Hence, the retaining wall model can be considered as a plane-strain model. This observation is consistent with the studies of Yang and Tang [30].

3.2 *Determination of Lateral Earth Pressures*

The following three cases are considered for the study for each combination of backfill:

1. Retaining wall is at at-rest condition (Ko)
2. Retaining wall moves out of the backfill i.e. active earth pressure condition (Ka)
3. Retaining wall moves into the backfill i.e. passive earth pressure condition (Kp).

Thus, a total of 30 laboratory model tests are conducted in this study to study the behaviour of sand-BDM backfill under three different modes of wall movement. The lateral pressures exerted on the back of the retaining wall are recorded during each test. It may be noted that the earth pressure values will differ depending on whether the pressure by the hydraulic jack is applied at the top or bottom of the retaining wall. In both active and passive cases, the model retaining wall in the present study is allowed to rotate about its top. The variation of lateral earth pressures for at-rest, active and passive earth pressure conditions are shown in Fig. 7a–c. The earth pressures are also compared with the classical theories of Jaky [11] and Coulomb [3].

From the above figures, it can be understood that experimental values are not matching with theoretical values. Addition of BDM increased at-rest earth pressure at all wall positions. This is because lower part of backfill received more compaction than upper part and addition of BDM also densified soils. In active case, wall movement is away from backfill generated through the backward movement of hydraulic jack. This also facilitates wall to rotate about its base point for 1–2°. It is observed that earth pressures recorded by top sensor during this active movement is in range of 0–1 kPa for every wall position. Middle sensor readings are between 1 and 2 kPa for all wall positions, whereas bottom sensor exhibited more horizontal stresses than other pressure cells. Experimental pressure readings are in between 4 and 6 kPa at all positions. Pressure readings are in contrast with theoretical values, both Jaky's theory and Coulomb's theory either under predicted or over-predicted the experimental values.

In case of passive pressures, it is observed that since the wall is made to rotate about its bottom, greater magnitude of earth pressure is observed near the top of the wall and it decreased along the depth of the wall. This observation is somewhat not matching with the established earth pressure theories of Jaky and Coulomb. The reason may be attributed to the fact that in this study, the since the wall is made to rotate about its top and the load from the hydraulic jack is applied at the top of the retaining wall, the passive earth pressure is greater near the top of the wall. This observation is consistent with that of Dave and Dasaka [4] and Yang and Tang [30]. Hence, it may be stated that the development of maximum pressure depends on whether the retaining wall is rotated about its top or bottom into the backfill soil. In passive case, greatest earth pressures occur at 1/3 height of wall from top which is in contrast with theoretical values.

Shear strength mobilization is increased with addition of BDM particles to existing soils among the backfill. The experimental results indicate that the earth pressures are not enhanced significantly by the inclusion of BDM to soil, which suggests that BDM can be used as an effective light-weight backfill. The optimum pressure is obtained on mixing 20% BDM with soil. From the results, it is clearly observed that the experimental findings vary significantly with the theoretical values. From the triaxial tests, it is noticed that the angle of internal friction increases with addition of BDM and the optimum is reached at 20% BDM. The backfill is assumed to be homogenous where vertical and lateral pressures remain constant. But in this case the backfill soil is mixed with BDM particles rendering it to be non-homogenous. The strength mobilization of backfill depends on the density and the angle of internal friction, which again is dependent on the compaction of soil. Shear strength was obtained on mixing 20% BDM with sand, the results were not consistent in case of earth pressures. Maximum earth pressure was obtained for 30% replacement of BDM, which was quite evident since the unit weight of backfill also increased with addition of BDM. In passive state, the strength exhibited by soil due to its arrangement is dominated by wall weight and horizontal pressure acting on it. Thus, full soil strength

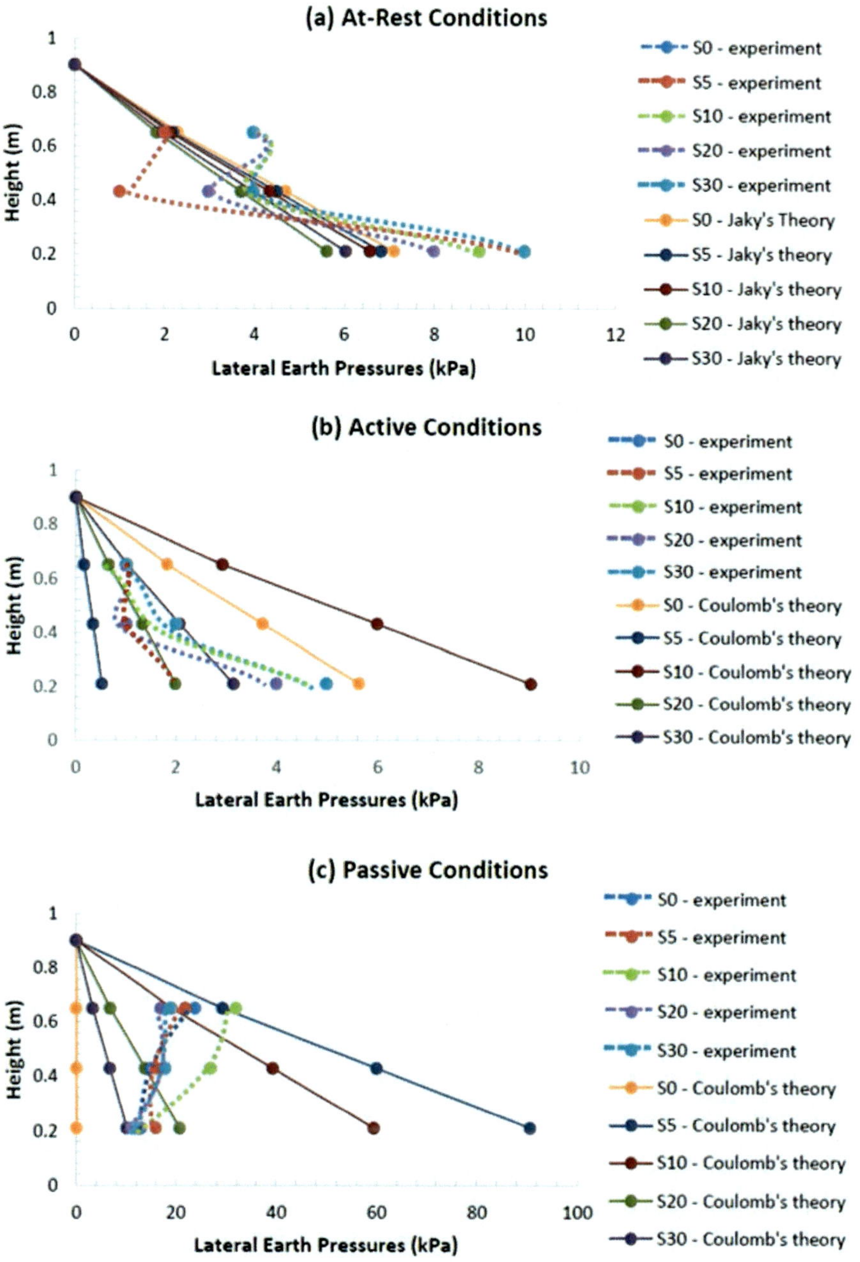

Fig. 7 Variation of lateral earth pressures for S-BDM for **a** at-rest, **b** active and **c** passive earth pressure conditions

is mobilized in this condition in all directions. Pressure envelops are developed along the height of wall, assuming zero earth pressures at top of backfill and no surcharge is acting on the backfill.

4 Finite Element Modelling of Retaining Wall

Large scale tests are generally expensive, and needs added labor cost that makes it difficult to carry out a wide range of parametric studies. In this regard, numerical analysis serves as a cost-effective measure that helps in better understanding the experimental results. A numerical model is a mathematical representation to a real physical problem. Although the most common methods of analysis for geosynthetic reinforced soil wall structures are based on simple limit-equilibrium methods, these methods are limited in their ability to predict stresses, forces and boundary reactions at working loads and offer no information on deformations and strains. Finite-element analysis of geotechnical structures requires an accurate model of the stress–strain behavior of the soil. This method is carried out to analyze behavior of soil at field conditions, thus giving an approximate insight on response of soil stratum under response of external loads.

In the present study, the cantilever retaining wall is modelled using commercially available finite element software PLAXIS 2D, which uses force-based or displacement-based methods. The force-based analysis is carried out by incorporating Hardening Soil (HS) model. The cantilever retaining wall, as used in the experimental study, is modelled using a plane strain model, with 0 displacements and strains in the z-direction. The retaining wall with different kinds of backfills with varying percentages of BDM is analyzed for at-rest, active and passive earth pressure conditions. Table 1 presents the parameters used in finite element modelling of the retaining wall.

The model is meshed with very fine global coarseness with global refinement. Since output is considerably affected by the number of generated elements, a convergence study is carried out by varying the number of mesh elements. Table 2 provides number of elements and nodes generated during meshing. Figure 8a and b provide the FE retaining wall and de-formed meshes of the retaining wall.

In Plaxis 2D, ΣMstage is a load intensity indicator which starts from 0 and reaches a maximum value of 1 at the end of the calculation phase. Depending on problem complexity, this value should be defined. Generally, ΣM-stage is defined as 1 which means load applies in normal way. Calculation of stresses in the initial phase is based on K0 procedure. The construction is carried out in the following 2 phases:

Stage-1: Generate initial stresses in the form of gravity loading.
Stage-2: Activate point load.

Figure 9a–c shows the variation of earth pressures of S-BDM with height of wall for at-rest, active and passive earth pressure conditions. It is observed that the earth pressures for all positions of retaining wall agrees well with the experimental result.

Table 1 Parameters used in PLAXIS 2D for retaining wall analysis

Element	Values
(1) Plate element	
Axial stiffness EA (kN/m)	1.25×10^7
Bending stiffness EI (kN m^2/m) Weight (w) (kN/m/m) Poisson's ratio (υ)	1.15×10^5 8.2 0.15
(2) Hinge element	Free to rotate at top
(3) Interface element $R_{\text{inter (S-BDM)}}$ Interface Strength	0.8 Rigid
(4) Loading position	0.75 m from base (monotonic-horizontal)
(5) Boundary conditions Wall (Plate element) Tank Soil	Rigid, non-yielding Fixed on all sides Homogenous, isotropic
(6) Test conditions	Dry backfill

Table 2 Number of elements and nodes generated with variation of mesh coarseness

Mesh coarseness	Elements generated	Nodes generated
Very coarse	211	1979
Coarse	472	4221
Medium	749	6583
Fine	1679	14,311
Very fine	3105	26,015

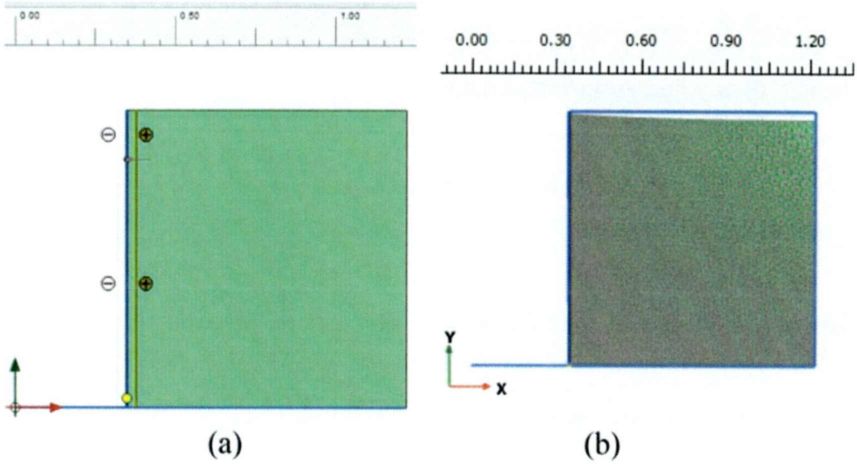

Fig. 8 **a** FE retaining wall model, **b** deformed mesh at K_0 condition

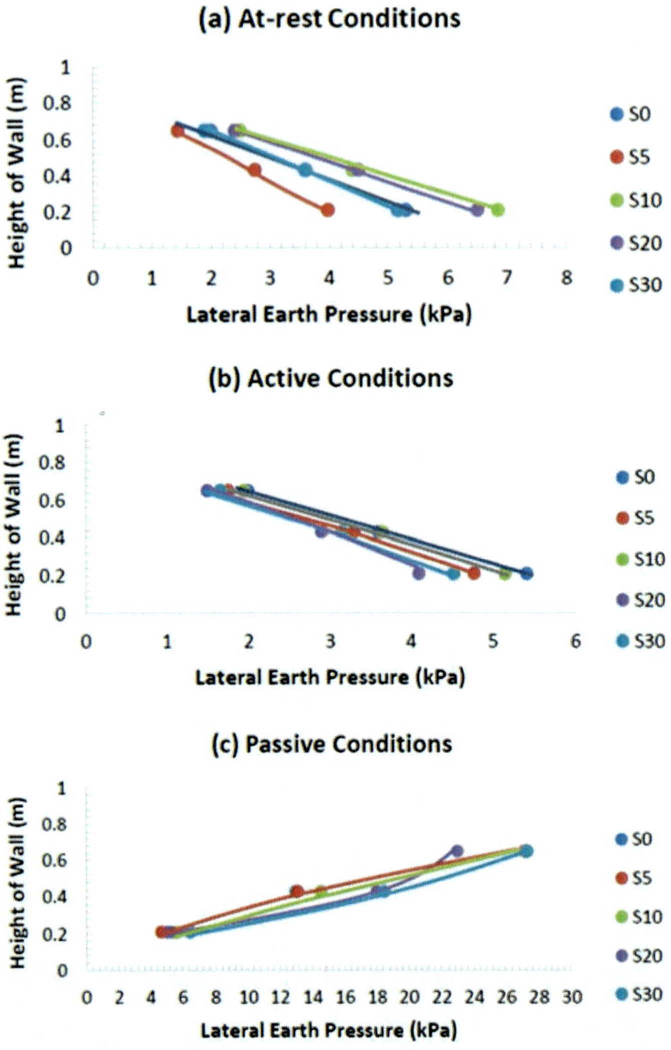

Fig. 9 Finite element results for **a** at-rest earth, **b** active and **c** passive pressure conditions

The addition of 20% BDM increases the earth pressures by about 18–22% at the bottom of the wall. This increase is not so significant towards the top of the wall.

5 Parametric Studies

To investigate the practical significance of utilizing BDM as a partial replacement of backfill soil, an example problem is simulated using the finite element software Plaxis 2018 v2. A 7 m high retaining wall with a backfill width of 6 m is considered

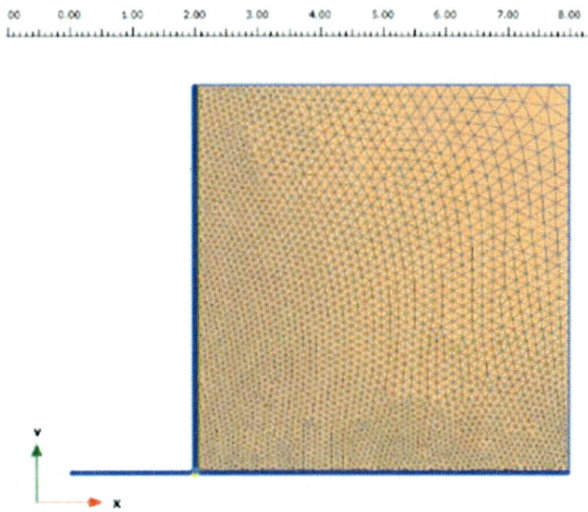

Fig. 10 Finite element model for 7 m high retaining wall

for the present analysis. A plane strain model is used for modelling the retaining wall. 15 noded triangular elements are used to discretize the model into 5313 elements and 43,647 nodes with very fine meshing. The geotechnical properties of sand and BDM are same as presented in Sect. 2. The finite element model with the meshing is provided in Fig. 10. The variation of at-rest, active and passive earth pressures for sand blended with different percentages of BDM content are plotted in Fig. 11a–c.

Hardening Soil (HS) model, which is based on Duncan–Chang Theory, is adopted for modelling stress–strain responses of soil. From Fig. 11, it is observed that the earth pressures in all three cases increased significantly with increase in wall height. The at-rest earth pressure varies between 54 and 59 kPa for sand-BDM blends. The range of earth pressures observed for sand-BDM in active case are in the range of 47–60 kPa. Similarly, for passive earth pressure conditions, the range is within 112–116 kPa for sand-BDM. Addition of BDM increases the earth pressures to a small extent; however, this rate of increase lies between 3 and 20% as compared to un-blended soils for all three cases. The maximum increase in earth pressure due to inclusion of BDM is observed for active cases, while for passive cases, inclusion of BDM does not significantly alter its values. Hence, it can be concluded that the inclusion of BDM does not significantly increase the earth pressures.

6 Conclusions

Proper disposal of the humungous volume of construction and demolition waste is one of the major issues in the today's world. For achieving sustain-ability and reducing carbon footprint, CDW is proposed to be reused in engineering applications. The present research work deals with the scopes of utilizing virgin BDM as

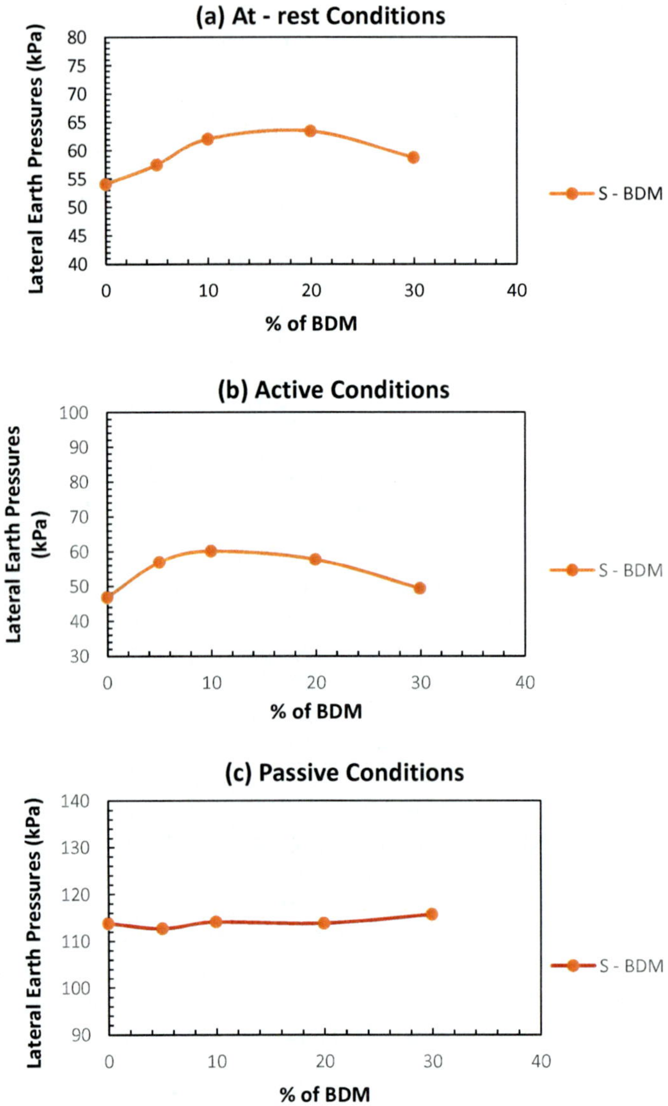

Fig. 11 Variation of **a** at-rest, **b** active and **c** passive earth pressures with BDM content

partial replacement for backfill soils of retaining walls. Locally available river sand is blended with 0–30% of BDM to improve the performance of the backfill soil. A numerical model is also developed to validate the experimental results. The major findings of the study on laboratory scale model retaining wall may be highlighted as follows:

- The experimentally determined at-rest earth pressures along the height of wall, is a little higher compared to the values obtained from Jaky's theory. The at-rest and active earth pressures increase with depth similar to established theories.
- For passive earth pressure conditions, since the wall is made to rotate about its top and the load from the hydraulic jack is applied at the top of the retaining wall, the earth pressure is greater near the top of the wall and decreased with depth, in contrast to classical Coulomb's earth pressure theory.
- For passive earth pressure conditions, Coulomb's theory under-predicts the earth pressure values for S-BDM. Although the passive pressures as calculated by Coulomb's theory falls in range towards the top part of the retaining wall, the theory significantly over-estimates the earth pressures towards the bottom of the wall. The reason is in the present study, the pressure is applied at the top of the retaining wall and Coulomb's theory does not consider the effect of the point of application of load.
- Since the present study considers rotation about top, in passive case, the point of application of earth pressures lies within 1/3 height of wall from top. This is in contrast with classical theories which do not differentiate between the point of application of pressure.
- Shear strength mobilization is increased with addition of BDM particles to existing soils among the backfill. The experimental results indicate that the earth pressures are not enhanced significantly by the inclusion of BDM to red soil, which suggests that BDM can be used as an effective light-weight backfill. The optimum pressure is obtained on mixing 20% BDM with red soil.
- A parametric study carried out with a 7 m high retaining wall shows that the earth pressures are not significantly increased by the inclusion of BDM.
- The finite element results for different wall positions show a good agreement with the experimental observations and the trend is also satisfactory.
- Overall, the proposed approach of partially replacing soil with BDM as a backfill material for retaining walls exhibits promising features and wide applicability for the analysis and design of geotechnical earth retaining structures. It may help to develop confidence to the practicing engineers in handling different types of wastes generated from construction industry and the proposed methodology may be adopted in practice for environment friendly and economic design of backfill for different earth structures.
- While the results presented and comments made in this thesis are based on specific soil properties and conditions mentioned, the same methodology can still be applied to any problem of geotechnical interest.

Funding

This work is funded by the DST-SERB, Govt. of India, Early Career Research Award (Project ID: ECR/2016/000522).

References

1. Akhtar A, Sarmah AK (2018) Construction and demolition generation and properties of recycled aggregate concrete: a global perspective. J Cleaner Prod 186:262–281
2. BMTPC Guidelines for Utilization of C&D Waste (for Construction of Dwelling Units and related Infrastructure in various Housing Schemes of the Government), Ministry of Housing & Urban Affairs, Government of India, 2017
3. Coulomb CA (1776) Essai sur une application des regles des maximis et minimis a quelques problemes de statique relatifs a l'architecture. Memoires de l'Academie Royale pres Divers Savants 7
4. Dave TN, Dasaka SM (2012) Transition of earth pressure on rigid retaining walls subjected to surcharge loading. Tunn Undergr Constr SP 242:836–844
5. Edil TB, Bosscher PJ (1994) Engineering properties of tire chips and sand mixtures. Geotech Test J 17(4):453–464
6. Fang Y, Ishibashi I (1986) Static Earth pressures with various wall movements. J Geotech Eng 112(3):317–333
7. Garside M (2020) Major countries in worldwide cement production from 2015–2019. Statista International
8. Hasan U, Chegenizadeh A, Budihardjo MA, Nikraz H (2016) Experimental evaluation of construction waste and ground granulated blast furnace slag as alternative soil stabilisers. J Geotech Geol Eng 34(6):1707–1722
9. Ichihara M, Matsuzawa H (1973) Earth pressure during earthquake. Soils Found 13(4):75–86
10. Jain S, Singhal S, Jain NK (2018) Construction and demolition waste (C&DW) in India: generation rate and implications of C&DW recycling. Int J Construct Manage. https://doi.org/10.1080/15623599.2018.1523300
11. Jaky J (1948) Pressure in silos. In: 2nd international conference on soil mechanics and foundation engineering, vol 1. London, pp 103–107
12. Jayatheja M, GuhaRay A, Suluguru AK, Anand A, Kar A (2018) Performance of cohesionless soil partially replaced with building derived materials as a foundation material under static loading conditions. Int J Geotech Eng, Taylor and Francis 1–10. https://doi.org/10.1080/19386362.2018.1543791
13. Limbachiya MC, Marrocchino E, Koulouris A (2007) Chemical–mineralogical characterisation of coarse recycled concrete aggregate. Waste Manage 27(2):201–208
14. Lin Y-L, Yang G-L, Li Y, Zhao L-h (2010) Engineering behaviors of gabion retaining wall based on laboratory test. J Central South Univ Technol, Springer, 17:1351–1356
15. Melton JS (2015) Recycled base aggregates in pavement applications—Part III of VI (AWI051611). Sustainable geotechnical applications: webinar hosted by ASCE
16. Meyer C (2002)Concrete and sustainable development. Special publication ACI 206, Concrete materials science to application—a tribute to Surendra P. Shah. American Concrete Institute, Farmington Hills, MI, pp 1–12
17. Meyer C (2004) Concrete materials and sustainable development in the USA. Struct Eng Int
18. NBO Report (2014) Building material prices and wages of labor—a statistical compendium, national buildings organisation, ministry of housing and urban development alleviation, government of India
19. Rao MC, Bhattacharya SK, Barai SV (2011) Influence of field recycled coarse aggregate on properties of concrete. Mater Struct RILEM 44:205–220
20. Roscoe KH, Schofield AN, Wroth CP (1958) On the yielding of soils. Géotechnique 8(1):22–53
21. Silva RV, Brito J, Dhir RK (2016) Performance of cementitious renderings and masonry mortars containing recycled aggregates from construction and demolition wastes. Construct Build Mater, Elsevier 105:400–415
22. Suluguru AK, Jayatheja M, Kar A, GuhaRay A, Surana SR, James N (2017) Experimental studies on the microstructural, physical and chemical characteristics of building derived materials to assess their suitability in ground improvement. Construct Build Mater, Elsevier 156:921–932

23. Suluguru AK, Surana SR, GuhaRay A, Kar A, Jayatheja M (2018) Experimental investigations on building derived materials in chemically aggressive environment as a partial replacement of soil in geotechnical applications. Geotech Geol Eng, Springer Netherlands. https://doi.org/10.1007/s10706-018-0662-0,37(2),947-963
24. Terzaghi K (1934) Large retaining wall tests. Engineering News Record
25. Terzaghi K (1936) Proceedings of first International Conference in International Society of Soil Mechanics and Foundation Engineering (Harvard)
26. The Hindu, Construction waste recycling plants soon (2017) Available at: http://www.thehindu.com/todays-paper/tp-national/tp-telangana/work-on-recyclingplants-to-start-soon/article20007457.ece. Last accessed on 09.11.17
27. USEPA (2015) Lifecycle challenge competition seeks new ideas to reduce construction and demolition debris. https://www3.epa.gov/region9/waste/solid/construction/. Last accessed on 11 Feb 2016
28. USEPA (2015) EPA advancing sustainable materials management: facts and figures 2013 8 186. https://doi.org/EPA530-R-15-002. Last accessed 10 July 2020
29. USEPA (2019) Advancing sustainable materials management: 2017 fact sheet. Last accessed 10 July 2020
30. Yang M, Tang X (2017) Rigid retaining walls with narrow cohesionless backfills under various wall movement modes. Int J Geomech 17(11)

Development of Region-Specific New Generation Attenuation Relations for North India Using Artificial Neural Networks

He Huang, R. Ramkrishnan, Sreevalsa Kolathayar, Ankit Garg, and Jitendra Singh Yadav

Abstract Present study focuses on developing region-specific New Generation Ground Motion Prediction Models using Artificial intelligence technique for North India purely based on a measured ground motion data from specific region. Simple single hidden layered feed forward multilayer perceptron networks with back-propagation learning algorithm are used. A total of 280 data points of recorded strong motion data from the Kangra and Uttar Pradesh (UP) arrays, made available by the Program for Excellence in Strong Motion Studies (PESMOS), were used to train these networks. The first model predicts Moment Magnitude for a given Hypocentral Distance and Peak Ground Acceleration. The second model predicts Peak Ground Acceleration (PGA) for a given Hypocentral Distance (HPD) and Moment Magnitude (MM). Performance analysis, Uncertainty analysis and analysis of interactive effects have been done to test the reliability of the generated models. Optimization analysis was also performed to predict possible inputs of the models for a given set of outputs. Models have performed reasonably well for the given amount of non-linearity in the data.

Keywords Ground motion · GMPE · Attenuation relationships · PGA · ANN · Seismic hazard analysis

H. Huang · A. Garg (✉)
Guangdong Engineering Center for Structure Safety and Health Monitoring, Shantou University, Shantou, China
e-mail: ankit@stu.edu.cn

R. Ramkrishnan
Department of Civil Engineering, Amrita School of Engineering, Amrita Vishwa Vidyapeetham, Coimbatore 641112, India

S. Kolathayar
Department of Civil Engineering, National Institute of Technology, Karnataka, India

J. S. Yadav
Department of Civil Engineering, National Institute of Technology Hamirpur, Hamirpur, H.P, India
e-mail: jitendershine@gmail.com

A. Garg et al. (eds.), *Proceedings of the 1st Indo-China Research Series in Geotechnical and Geoenvironmental Engineering*, Lecture Notes in Civil Engineering 123,
https://doi.org/10.1007/978-981-33-4324-5_6

List of Symbols

PESMOS	Program for excellence in strong motion studies
PGA	Peak ground acceleration
HPD	Hypocentral distance
MM	Moment magnitude
GMPE	Ground motion prediction equations
ANN	Artificial neural networks
NGA	New generation attenuation
MLP	Multi-layer perceptron
MFT	Main frontal thrust
MBT	Main boundary thrust
MCT	Main central thrust
NIC	National informatics center
SMA	Strong motion accelerograph
CESMD	Center of engineering strong motion data
VDC	Virtual data centre
COSMOS	Consortium of organizations for strong motion observation systems
BFGS	Broyden-Fletcher-Goldfarb-Shanno
RBF	Radial basis function
TRP	Training performance
TP	Testing performance

1 Introduction

The Himalayas has witnessed large magnitude earthquakes from ancient times and is considered highly vulnerable to earthquake hazards [1–4]. Seismic zonation map of India has a major disadvantage that it was developed based on previous seismic events without any region specific measured data that captures both space and time effect. Some studies [5–9] have attempted to capture seismic hazard analysis of India, however a region specific attenuation relations has been rarely captured [3, 9]. Recently, region-specific attenuation relations [9] were developed. Ground Motion Prediction Equations (GMPE) adopted in seismic hazard analysis are mainly for regions with similar identical tectonic settings. In addition, many of these equations were based on limited measured ground motion data along with simulated data [2, 10]. Hence, there is a need to develop a GMPE based equations based on significant measured ground motion data for specific north India region.

As per Kumar et al. [11], GMPEs are empirical relations that are established based on various parameters. These parameters include magnitude, distance from source to site, type of fault, depth of focal, source parameters, propagation path and site parameters. It is well known that Peak Ground Acceleration (PGA) is generally adopted to quantify extent of ground shaking. GMPEs can be useful to establish seismic hazard maps. These maps can be utilized for developing preliminary guidelines or analysis for building codes. Artificial Neural Networks (ANN) were used to predict PGA in many recent studies [12–16]. ANN perform better than conventional modelling techniques due to their ability to learn without any assumptions regarding the problem [17]. ANN can develop solutions (i.e. find functional relations) efficiently for complex and non-linear interactions between variables in data [14].

In past, there has been attempt to develop Region-specific New Generation Attenuation Relations (NGA) depending on the availability of dataset [9, 18–21]. A larger dataset for the Himalayas is available [9] that contains parameters required for establishing reliable GMPE (as mentioned in previous sub-section). Present study proposes a new Ground Motion Prediction Model which is developed purely based on a measured data of actual recorded ground motion of Himalayas in North India using ANN. The developed models are not specific equations but are programs or codes consisting of ANN to predict the outputs. Interactive effects of input parameters are also considered in the generated models. Here, multi-layer perceptron (MLP) networks were used to develop efficient models based on their performances. Two different models are developed for efficient use in seismic hazard analysis. First one requires HPD and PGA as inputs to predict Moment Magnitude (MM) as output. Next one requires HPD and MM as inputs to predict PGA as output. Uncertainty and optimization analysis were performed on the models. The codes of the models are given in the appendix, these can be used as equations/black boxes to predict the output by giving the appropriate inputs (refer Table 3).

2 Methodology

2.1 Indian Strong Motion Database and the Himalayan Seismicity

Northern Himalayan region has major tectonic features such as Main Frontal Thrust (MFT), Main Boundary Thrust (MBT) and the Main Central Thrust (MCT). These features are mainly formed due to collision of Indo-Eurasian tectonic plates [22]. The ground motion data recorded at various stations under National Seismological Center is transferred to the National Informatics Center (NIC) Delhi, and to IIT Roorkee for further processing. The processed data is made available in PESMOS. The dataset consists of detailed parameters required for establishing GMPEs as mentioned before. Figure 1 shows the geographical locations of the recording stations and the earthquakes considered in the study, on a seismotectonic map showing the major faults in the region.

Data from 2005 to 2016 was directly obtained from PESMOS website, Center of Engineering Strong Motion Data (CESMD) (http://strongmotioncenter.org) and the Virtual Data Centre (VDC) of the Consortium of Organizations for Strong Motion

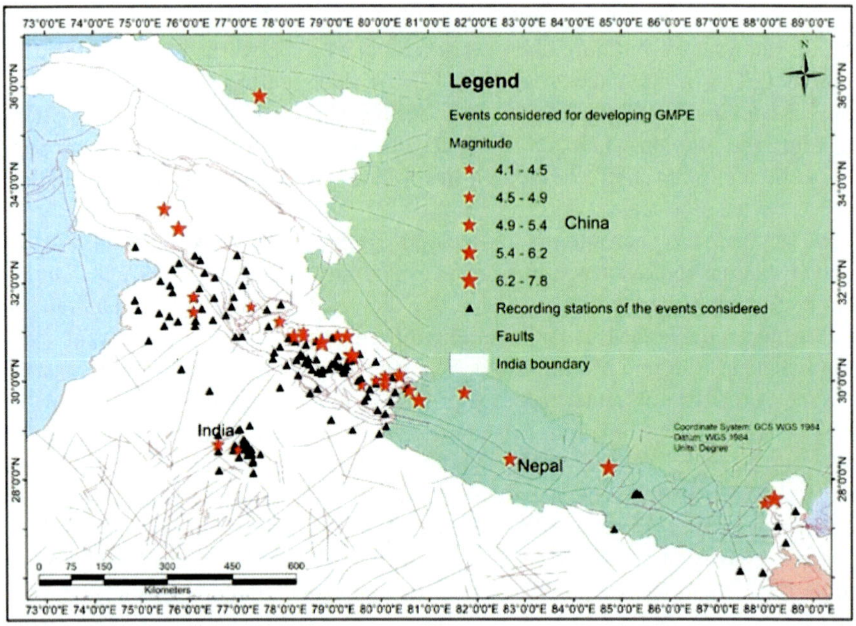

Fig. 1 Seismotectonic map of the region showing seismic sources, recording stations and events considered

Observation Systems (COSMOS) (http://strongmotioncenter.org/vdc). The ground motion data corresponds to both soil and rock sites. A graph of the compiled data showing the log Hypocentral distance vs magnitude is plotted in Fig. 2b. Details of dataset utilized in this study is summarized in Ramkrishnan et al. [9].

2.2 *Modelling Using Artificial Neural Networks*

ANN was found in 1940's by McCulloch and his colleagues. It was inspired from the biological neural networks in the human brain and is now used as a popular statistical learning model [16, 23]. ANN are highly popular for their use in Artificial Intelligence, which proves its relevance to develop models for complex and non-linear data [17]. The networks used in this study contain a single hidden layer with an input and an output layer. Each layer consists of neuron(s) commonly addressed as node(s) (refer Fig. 3). The nodes are interconnected among the layers, where each link (connection) is associated with a weight. Weights (W_{ij} and W_{jk}), which represent interconnection strength, describe the importance of a specific link. Using lower number of nodes (in hidden layer) lowers the performance of the network, on the other hand, higher number of nodes (in hidden layer) will make the network lose its ability for generalization (over learning takes place). Figure 3a shows a typical

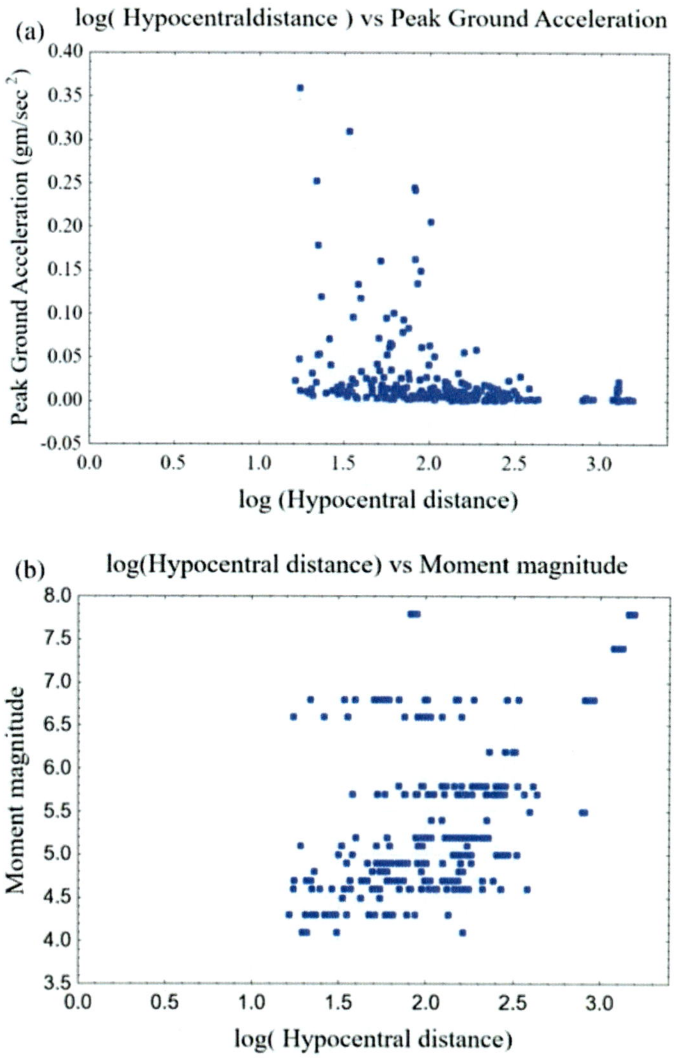

Fig. 2 **a** PGA-distance chart; **b** magnitude-distance chart

feed forward single hidden layer neural network. Figure 3b shows the schematic representation of a neuron or a node in the network. The networks were trained using software Statistica version 12. There are two different network types available in Statistica namely, Multilayer perceptron and Radial basis functions. Both of them were used to model the data and the best performing models were chosen.

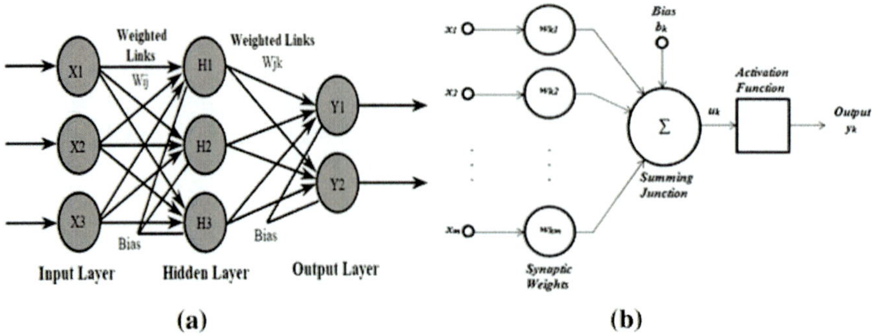

Fig. 3 **a** Schematic representation of an Artificial Neural Network, **b** schematic representation of a neuron in network

3 Multilayer Perceptron Networks

In multilayer perceptron (MLP) type of networks, all the inputs are summed at the summing junction after being multiplied with respective weights associated to links (refer Fig. 3). Later, a bias is added to this sum and the total sum is transformed through an activation function (f (.)). Rel U and sigmoid are two most commonly used activation functions. Same procedure is followed once the transformed sum is carried to the next layer. Mathematically the equation after passing through the hidden layer is given in Eq. 1. Here, Eq. 2 represents the mathematical form of process in the network after data passes through the output layer. In the beginning, all the weights and biases are chosen randomly. Once the result is obtained for a specific input, it is compared with the given value of respective output from the training data and mean square error is calculated. Then, based on that error, the weights and biases are adjusted and updated using back propagation algorithm. This process is called training of the network. Difference between the predicted and given value is called the cost value or error value and it is desired to have a negligible magnitude. Until the least desired square of error value is obtained, the training process is repeated for all the training data. After several repetitions, the network arrives with suitable weights and biases. Referring to training and testing performances of various networks containing different configurations, an efficient network is chosen (Table 1). The notation of an MLP network is as follows. MLP 2-29-1 implies that network has 2 neurons in input layer, 29 neurons in hidden layer and one neuron in output layer.

$$Z_k = f\left[\sum_{i=1}^{n}(W_{ij}.X_i + b_1)\right] \tag{1}$$

Z_k is the output from hidden layer, f (.) is the activation function between input and hidden layers and n is the number of nodes in the input layer. b_1 is the hidden layer bias. W_{ij} are the weights of links between input and hidden layer.

Table 1 List of events considered and number of stations at which ground motion was recorded

No.	Depth	Date	Location	Lat	Long	Mw	No of stations
1	10	Oct 19 1991	Uttarakashi	30.7	78.7	6.8	13
2	15	Mar 28 1999	Chamoli	30.5	79.4	6.6	11
3	25.7	Dec 14 2005	Chamoli	30.9	79.3	5.2	8
4	20.3	Nov 25 2007	Delhi Haryana Border	28.6	77.0	4.3	10
5	15	Aug 19 2008	Pithorgarh-Uttarakhand	30.1	80.1	4.3	4
6	10	Sep 4 2008	Uttarakhand Tibet Border	30.1	80.4	5.1	7
7	10	Oct 21 2008	Kullu-Himachal Pradesh	31.5	77.3	4.5	3
8	13	Sep 21 2009	Uttarkashi	30.9	79.1	4.7	12
9	8	Sep 21 2009	Bhutan	27.3	91.5	6.2	4
10	15	Oct 3 2009	Bageshwar	30.0	79.9	4.3	3
11	2	Feb 22 2010	Bageshwar	30.0	80.1	4.7	6
12	29	Mar 14 2010	Himachal Punjab Border	31.7	76.1	4.6	13
13	10	May 1 2010	Bageshwar	29.9	80.1	4.6	7
14	43	May 28 2010	Himachal	31.2	77.9	4.8	4
15	10	Jul 10 2010	Almora Uttarakhand	29.9	79.6	4.1	4
16	5	Jan 18 2011	Southwestern Pakistan	28.9	64.0	7.4	12
17	10	Apr 4 2011	Nepal-India Border	29.6	80.8	5.7	24
18	26	Jun 3 2011	Sikkim-Nepal Border	27.5	88.0	4.9	4
19	12	Jun 20 2011	Chamoli-Uttarakhand	30.5	79.4	4.6	13
20	10	Sep 18 2011	Sikkim-Nepal Border	27.6	88.2	6.8	9

(continued)

Table 1 (continued)

No.	Depth	Date	Location	Lat	Long	Mw	No of stations
21	10	Feb 9 2012	Uttarakashi	30.9	78.2	5.0	5
22	14	Mar 5 2012	Haryana Delhi Border	28.7	76.6	4.9	21
23	10	Aug 23 2012	Nepal	28.4	82.7	5.0	3
24	12	Nov 27 2012	Uttarakashi	30.9	78.4	4.8	4
25	34	Jan 9 2013	Nepal	29.7	81.7	5.0	4
26	5	Feb 2 2013	Uttarakashi	31	78.4	4.3	3
27	46	Apr 16 2013	Iran Pakistan Border	28	62.1	7.8	6
28	15	May 1 2013	JK-HP Border	33.1	75.8	5.8	21
29	28	Aug 2 2013	JK-HP Border	33.5	75.5	5.4	3
30	10	Aug 29 2013	Himachal-Punjab-Border	31.4	76.1	4.7	8
31	80	Sep 20 2013	India China Border	35.8	77.5	5.5	4
32	8.2	Apr 25 2015	Gorkha, Nepal	28.2	84.7	7.8	5
33	10	Dec 1 2016	Nepal-India	29.8	80.6	5.2	20

$$Y_k = f'\left[\sum_{j=1}^{m} W_{jk}.f\left[\sum_{i=1}^{n}\left(W_{ij}.X_i + b_1\right)\right] + b_2\right] \quad (2)$$

Y_k is the output from the output layer itself. f' is the activation function between hidden and output layers, m is the number of nodes in hidden layer and b_2 is the bias of the output layer [12]. BFGS (Broyden-Fletcher-Goldfarb-Shanno) algorithm is used to train the MLP network in Statistica since it performs well than gradient descent algorithms but is memory intensive (Statistica 2018).

4 Radial Basis Functions

Radial basis function (RBF) networks were initially introduced by Broomhead and Lowe in 1988 [24]. RBF networks learn using a supervised training technique and consists of a single hidden layer with an activation function as Gaussian response curve. Training an RBF network essentially means to decide appropriate receptors and spreads of clusters for all nodes in the hidden layer along with updating weights for appropriate output (refer to Fig. 4). Generally, in most of the applications involving RBF networks, hidden layer is of high dimensionality (consists a greater number of neurons than input layer) compared to the input layer. Increase in dimensionality followed by transformation makes the task of differentiating easy among clusters.

Centre of the clusters are called the receptors. The transfer function of neuron (Gaussian function) indicates the influence of data points at the centre [12]. Receptors, can be selected randomly from training data. Receptors can be also be trained iteratively, or derived using techniques such as Max-Min algorithms and K-means [12, 25, 26]. From every receptor (i.e. centre of cluster), root mean square of a few nearest neighbors decides the value of sigma in Gaussian function in Eq. 3.

$$G||x_i - c_i|| = -e^{\sum_{i=1}^{m} \frac{-||x_i - c_i||}{2\sigma^2}} \tag{3}$$

Kohonen self-organizing maps were used here. After this cluster formation and deciding the receptors is completed, the weights between the hidden and output layer nodes are found by multiple regression in a supervised manner [12]. The output of the RBF network is given in Eq. 4.

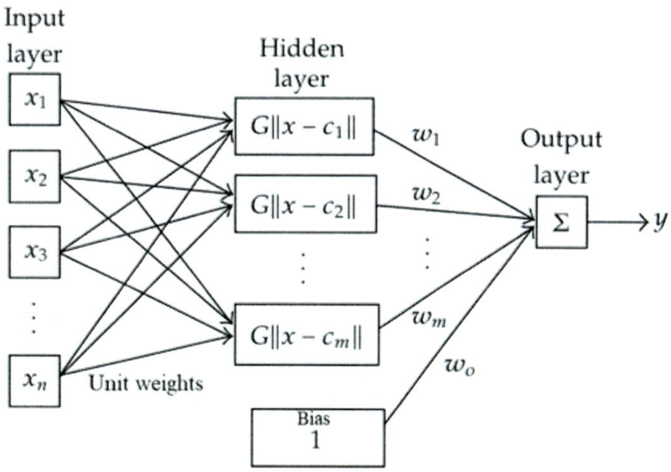

Fig. 4 Schematic representative diagram of RBF network (after Günaydin et al. [12])

Table 2 Parameters of artificial neural networks

Parameter	Values
Training data	80–85%
Testing data	10–15%
Validation data	0%
No. of hidden layers	1
No of neurons in a hidden layer	8–21
Target goal mean square error	10^{-5}
Minimum performance gradient	10^{-5}
Seeds to sample	1000

$$y = \sum_{i=1}^{m} w_i.G||x - c_i|| + w_o \tag{4}$$

The term w_i represents the weights of links between hidden and output layer, m is the number of hidden nodes (refer to Fig. 4), x is the input feature vector, $||x-c_i||$ represents the Euclidian distance between known data location x and prediction location c_i. Centre of the respective cluster or field is depicted by c.

The obtained result y is compared to known results in training data and mean square error is estimated. Based on the error, appropriate receptors and spreads of clusters are altered to minimize it. Various networks were trained and the network with best training, testing and validation performance was chosen. MLP 2-21-1 means that there are 2 nodes in input layer, 21 nodes in hidden layer and 1 node in output layer.

Parameters used for ANN

The parameters of training (for two models) have varied for those models in order to obtain best performances. Hypocentral distance was the constant input for all the models. Table 2 shows all the parameters used to train the networks.

5 Results and Discussions

Regression analysis using ANN was carried out to obtain two different models. The performances of the models were satisfactory for the given non-linearity of data. The new ground motion prediction equations are the models of ANN and are not in the form of an equation. These models are trained artificial neural networks, which were chosen based on their performances to predict the output. Thus, these networks are expressed in form of code in C++. Table 3 shows the details of the models. These models would be only valid in the northern region of India since the data used in the analysis are specific to that particular region.

Table 3 Details of newly predicted ground motion equation models

Model no.	Network Type	Input	Output	Activation function
1.	MLP 2-08-1	HPD and PGA	Moment magnitude	Exponential
2.	MLP 2-21-1	HPD and magnitude	PGA	Sigmoid

a. Performance analysis

All the models have been chosen after rigorous analyses which includes training and testing thousands of neural networks while varying different parameters such as training, testing and validation data. Training performance (TRP), testing performance (TP) and plots (obtained vs. predicted) are the measure for the wholesome performance of the models. Model 11 has a training performance of 84%, testing performance of 91% and R_2 value of 0.625. Model 2 has a training performance of 87%, testing performance of 92% and R_2 value of 0.80. Only R_2 values are not the appropriate measures to judge the performance since it measured by magnitude of distances from the line rather than seeing the actual error and number of points nearer. Thus, all three aspects (TRP, TP and R2) should be considered while choosing. Given a highly nonlinear data, which can be witnessed from Figs. 1 and 2, the models perform reasonably well. Moreover, to make the model simple, only 1 hidden layer has been used. Use of multiple hidden layers could achieve better results but would make the model complex and hard to perform various analyses to check its validity (Fig. 5).

b. Interactive effects of parameters

Hypocentral distance is the common input parameter in all the models. This is because it is the only parameter which can be determined on the field. Knowing the way how parameters affect each other in a model one can judge its reliability. If the learnt trends are valid physically then the model will be more reliable. Interactive effects can be easily understood by 3D plots for the models which have more than one input.

Model 1:

For a given hypocentral distance, PGA increases as moment magnitude increases (refer Fig. 6a). This is physically a valid phenomenon. Figure 6b shows moment magnitude at a particular PGA and Hypocentral distance. It shows lower PGA at higher distances, even at higher moment magnitudes. These show the physical validity of the model and hence make it reliable to some extent.

Model 2:

Figure 7 it shows an appropriate trend of PGA, as hypocentral distance increases PGA decreases. Also, majority of the plot depicts the physical phenomenon where PGA is high at places near to the hypocentre. The trend shown for lower moment magnitudes should be avoided since it deviates from physical behavior.

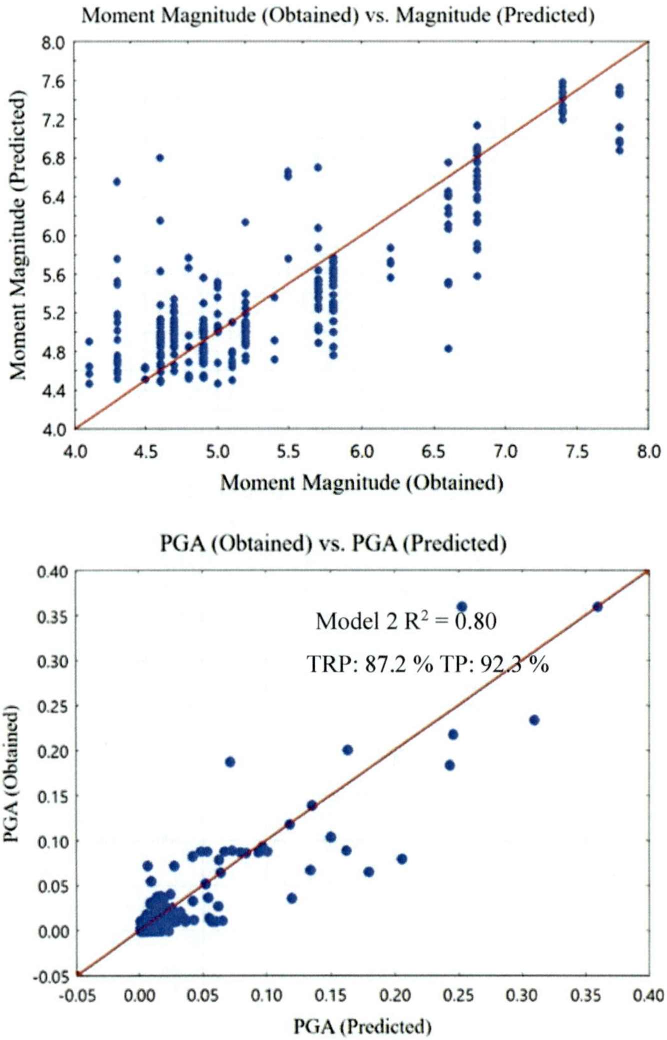

Fig. 5 Obtained versus predicted data plots for performance analysis

c. Uncertainty analysis

While using model 1, the output moment magnitude can fit into a log normal distribution. Even for model 2 an approximation of log normal distribution for PGA with low standard deviation can be used (refer Fig. 8a, b).

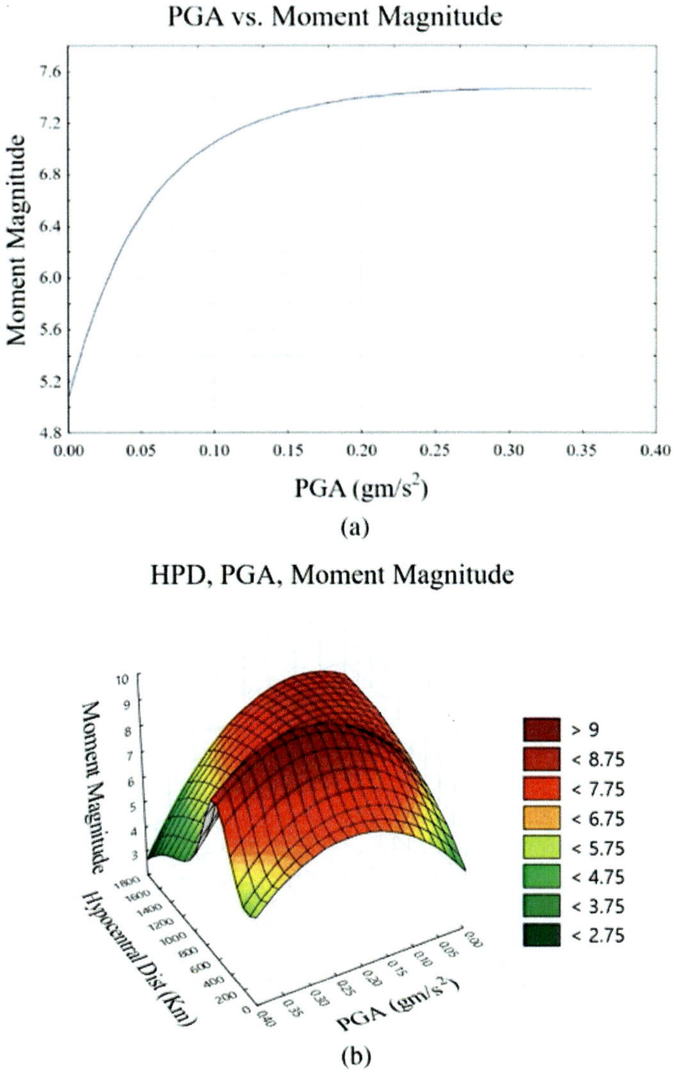

Fig. 6 Plots to analyze interactive effects for model 1

6 Optimization Analysis

Highest and lowest outputs were the targets in the optimization analysis, at which the appropriate inputs must be obtained. Simplex and exhaustive grid search algorithms were used for optimization. Design of few applications would require optimization analysis (Tables 4 and 5).

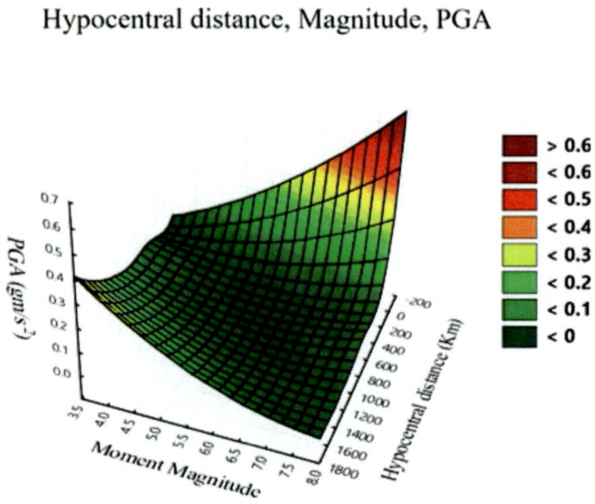

Fig. 7 Plot to analyze interactive effects of model 2

7 Conclusion

This study focuses on developing new generation region specific Ground Motion Prediction Model using ANN. Recorded strong motion data from particular region in North India (i.e., Kangra and Uttar Pradesh (UP) province arrays) were utilized for development of equations. All data from 2005 to 2016 was obtained from PESMOS, Center of Engineering Strong Motion Data and the Virtual Data Centre of the Consortium of Organizations for Strong Motion Observation Systems. In total, 33 earthquake events with moment magnitude ranging from 4.1 to 7.8; PGA from 0.001 to 0.36 g and hypo-central distance from 16 to 1560 km was considered in collected data [9]. Two different models with best performances were developed to predict PGA and MM after training several ANNs. Table 3 gives the information of inputs and output of the models. Both the models have performances more than 80%. Interactive effects of the models were analyzed to check their reliability. Uncertainty analysis was performed to observe the obtained distributions of the outputs in respective models. For the given non-linearity of the data models have performed reasonably well and ANN give better results when compared to response surface methodologies or any conventional modelling techniques. ANN using multiple hidden layers and hybrid algorithms might perform well. Both the models are reasonably efficient and reliable for their use in real world applications.

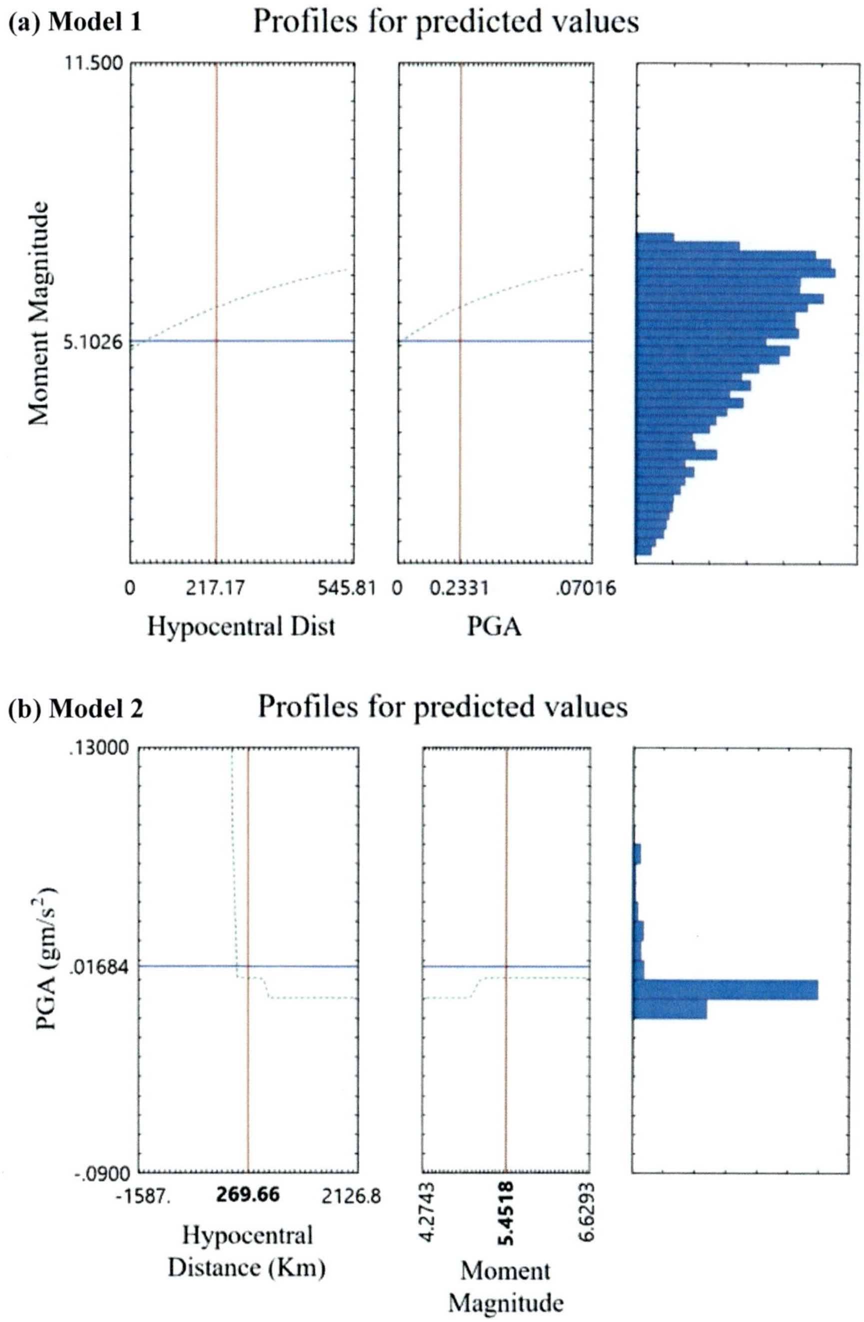

Fig. 8 Results of Uncertainty analysis

Table 4 Optimization analysis for model 1

Model 1	Input: HPD (Km)	Output: MM	Input: PGA (m/s^2)
Target: highest MM	480	7.7	0.263
Target: 6	237	6	0.025
Target: 5	154.6	5	0.004

Table 5 Optimization analysis for Model 2

Model 2	Input: HPD (km)	Input: MM	Output: PGA (m/s^2)
Target: highest PGA	22	7.09	0.36 g
Target: lowest PGA	664.68	4.04	0.0004 g
Target: 0.25 g	18.7	6.48	0.25

References

1. Khattri KN (1987) Great earthquakes, seismicity gaps and potential for earthquake disaster along the Himalaya plate boundary. Tectonophysics 138:79–92. https://doi.org/10.1016/0040-1951(87)90067-9
2. Kolathayar S, Sitharam TG, Vipin KS (2012) Spatial variation of seismicity parameters across India and adjoining areas. Nat Hazards 60:1365–1379. https://doi.org/10.1007/s11069-011-9898-1
3. Anbazhagan P, Kumar A, Sitharam TG (2013) Ground motion prediction equation considering combined dataset of recorded and simulated ground motions. Soil Dyn Earthq Eng 53:92–108. https://doi.org/10.1016/j.soildyn.2013.06.003
4. Gupta AK (2017) Response spectrum method in seismic analysis and design of structures
5. Shanker D, Sharma ML (1998) Estimation of seismic hazard parameters for the Himalayas and its vicinity from complete data files. Pure appl Geophys 152:267–279. https://doi.org/10.1007/s000240050154
6. Mahajan AK, Thakur VC, Sharma ML, Chauhan M (2010) Probabilistic seismic hazard map of NW Himalaya and its adjoining area, India. Nat Hazards 53:443–457. https://doi.org/10.1007/s11069-009-9439-3
7. Kolathayar S, Sitharam TG, Vipin KS (2012) Deterministic seismic hazard macrozonation of India. J Earth Syst Sci 121:1351–1364. https://doi.org/10.1007/s12040-012-0227-1
8. Sitharam TG, Kolathayar S (2013) Seismic hazard analysis of India using areal sources. J Asian Earth Sci 62:647–653. https://doi.org/10.1016/j.jseaes.2012.11.013
9. Ramkrishnan R, Sreevalsa K, Sitharam TG (2019) Development of new ground motion prediction equation for the North and Central Himalayas using recorded strong motion data. J Earthquake Eng 1–24
10. Yadav RBS, Koravos GC, Tsapanos TM, Vougiouka GE (2014) A probabilistic estimate of the most perceptible earthquake magnitudes in the NW Himalaya and adjoining regions. Pure Appl Geophys 172:197–212. https://doi.org/10.1007/s00024-014-0864-1
11. Kumar R, Choudhury D, Bhargava K (2016) Determination of blast-induced ground vibration equations for rocks using mechanical and geological properties. J Rock Mech Geotech Eng 8:341–349. https://doi.org/10.1016/j.jrmge.2015.10.009
12. Günaydin K, Günaydin A (2008) Peak ground acceleration prediction by artificial neural networks for northwestern Turkey. Math Probl Eng. https://doi.org/10.1155/2008/919420

13. Ahmad I, El Naggar MH, Khan AN (2008) Neural network based attenuation of strong motion peaks in Europe. J Earthq Eng 12:663–680. https://doi.org/10.1080/13632460701758570
14. Estrada GA, Baldani VL, De Oliveira D, Urquiaga S, Baldani JI (2013) Selection of phosphate-solubilizing diazotrophic Herbaspirillum and Burkholderia strains and their effect on rice crop yield and nutrient uptake. Plant andSoil 369:115–129. https://doi.org/10.1007/s11104-012-1550-7
15. Dhanya J, Gade M, Raghukanth STG (2017) Ground motion estimation during 25th April 2015 Nepal earthquake. Acta Geod Geophys 52:69–93. https://doi.org/10.1007/s40328-016-0170-8
16. Khosravikia F, Zeinali Y, Nagy Z, Clayton P, Rathje EM (2018) Neural network-based equations for predicting PGA and PGV in Texas, Oklahoma. Geotech Earthq Eng Soil Dyn 538–549
17. Maran JP, Sivakumar V, Thirugnanasambandham K, Sridhar R (2013) Artificial neural network and response surface methodology modeling in mass transfer parameters predictions during osmotic dehydration of Carica papaya L. Alexandria Eng J 52:507–516. https://doi.org/10.1016/j.aej.2013.06.007
18. Abrahamson N, Silva W (2008) Summary of the Abrahamson & Silva NGA ground-motion relations. Earthq Spectra 24:67–97. https://doi.org/10.1193/1.2924360
19. Campbell KW, Bozorgnia Y (2008) NGA ground motion model for the geometric mean horizontal component of PGA, PGV, PGD and 5% damped linear elastic response spectra for periods ranging from 0.01 to 10 s. Earthq Spectra 24:139–171. https://doi.org/10.1193/1.2857546
20. Kaklamanos J, Baise LG (2011) Model validations and comparisons of the Next Generation Attenuation of ground Motions (NGA-West) project. Bull Seismol Soc Am 101:160–175. https://doi.org/10.1785/0120100038
21. Idriss IM, Eeri HM (2014) An NGA-West2 empirical model for estimating the horizontal spectral values generated by shallow crustal earthquakes. Earthq Spectra 30:1155–1177
22. Parvez IA, Sutar AK, Mridula M, Mishra SK, Rai SS (2008) Coda Q estimates in the Andaman islands using local earthquakes. Pure Appl Geophys 165:1861–1878. https://doi.org/10.1007/s00024-008-0399-4
23. Perlovsky L (2001) Neural networks and intellect : using model-based concepts. Oxford University Press
24. Broomhead DS, Lowe D (1988) Radial basis functions, multi-variable functional interpolation and adaptive networks. Tech Rep 2:4148
25. Kohonen T, Kaski S, Somervuo P, Lagus K, Oja M (1997) Self-organizing map, 2nd edn. Springer, Berlin, Germany
26. Haykin S (1999) Neural networks: a comprehensive foundation, 3rd edn. Prentice-Hall, Englewood Cliffs, NJ, USA. 978-0131471399

The Effect of Sand Ratio on Suction and Swelling Pressure of Two Bentonite–Sand Mixtures

Ramakrishna Bag and Koteswaraarao Jadda

Abstract Bentonite–sand mixtures are chosen as potential barrier materials in various waste disposal system. This paper presents the effect of sand ratio on suction and swelling properties of two bentonite–sand mixtures. Constant volume swelling pressure tests on compacted bentonite–sand mixtures specimens and suction measurements on saturated specimens were carried out. The water holding capacity of the monovalent bentonite–sand mixtures were noted to be higher at lower suction range of 3 MPa, whereas at higher suction range the divalent bentonite–sand mixtures having greater water retention capacity. At a given suction, the increase in sand ratio leads to decreasing the water retention capacity of both the bentonite–sand mixtures. At a given sand ratio the swelling pressure exhibited by divalent bentonite–sand mixture was significantly higher as compared to monovalent bentonite–sand mixtures. Further the swelling pressure was noted to be decreased with an increase in sand ratio.

Keywords Bentonites · Sand · Sand ratio · Suction · Swelling pressure

1 Introduction

Due to low permeability, high swelling capacity, and retention properties, the compacted bentonites and bentonite–sand mixtures have been extensively in use for various geotechnical applications, such as sealing material in the galleries of high-level toxic waste disposal repositories [13]; and in vertical cutoff walls [6]. Understanding the water retention properties as well as swelling pressure behavior of barrier material is an essential aspect of the long-term stability of the barrier materials. The sand-bentonite mixture is low compressible, less susceptible to frost damage, and possess higher thermal conductivity than that of bentonite [5]. The addition of sand content to bentonite increases the maximum dry density of the bentonite–sand mixtures and decreases corresponding optimum water content, as

R. Bag (✉) · K. Jadda
Department of Civil and Environmental Engineering, IIT Patna, Patna 801106, India
e-mail: rkbag@iitp.ac.in

A. Garg et al. (eds.), *Proceedings of the 1st Indo-China Research Series in Geotechnical and Geoenvironmental Engineering*, Lecture Notes in Civil Engineering 123,
https://doi.org/10.1007/978-981-33-4324-5_7

well as decreases Atterberg limits [18]. Komine [10] studied the detailed investigation on the thermo-hydro-mechanical properties of the bentonite–sand mixtures and proposed various empirical relations for the hydraulic conductivity of backfill materials by incorporating various parameters such as bentonite content and dry density. Cui et al. [4] investigated swelling pressure and swelling deformation of GMZ bentonite–sand mixtures, consisting of different percentages of sand contents such as 10, 20, 30, 40, and 50%. Bag et al. [2] studied the effect of sand ratio on swelling pressure and hydraulic conductivity of an Indian bentonite–sand mixture.

Ravi and Rao [11] investigated the solute concentrations in micro-pore and macro-pore solutions of compacted bentonite–sand mixtures. The study concluded that the exchangeable cation concentrations in the micro-pore solution exceed the soluble salt-cation concentrations in the macro-pore solution of Barmer bentonite. Chen et al. [3] studied the compression, swelling, and rebound behavior of GMZ bentonite/additive mixture under coupled hydro-mechanical condition. The study results indicated that the volumetric deformation characteristics of the mixture largely depends on the initial condition, additive content, and type. Sun et al. [15] studied the swelling characteristics of GMZ bentonite and its mixtures with sand. The test result shows that, the relation between the void ratio and the swelling pressure of compacted GMZ bentonite–sand mixtures at full saturation is independent of the initial conditions such as the initial dry density and initial water content, and dependent on bentonite/sand ratio. Srikanth and Mishra [14] investigated the role of particle size of sand on the geotechnical characteristics of sand-bentonite mixtures. The results concluded that irrespective of the sand particle size, mixtures with bentonite content less than 20% showed a general lack of noticeable swelling.

Agus et al. [1] studied the swelling pressure-suction relationship of heavily compacted bentonite–sand mixtures. The study concluded that compacted specimens did not exhibit any collapse upon suction decrease and exhibited maximum swelling pressures at zero-equilibrium suction. Gatabin et al. [7] studied the competing effects of volume change and water uptake on the water retention behavior of a compacted MX-80 bentonite/sand mixture. The study concluded that the water retention capacity was higher under free swelling conditions than it was under prevented swelling. Ravi and Rao [12] investigated the influence of infiltration of sodium chloride solutions on soil–water characteristics curves (SWCC) of compacted bentonite–sand specimens. The study concluded that the infiltration of sodium chloride solutions has less effect on the microstructure and SWCC relations of bentonite–sand specimens compacted at higher dry densities.

In the current study, a series of constant volume swelling pressure tests were conducted on two different compacted bentonite–sand mixtures by varying sand content. The suction measurements were carried out on bentonite–sand slurries at different sand ratios.

2 Materials and Methodology

2.1 Materials

Commercially available two different types of bentonites were used in the present study. Based on cation exchange capacity, these bentonites are classified as divalent and monovalent bentonite. The bentonites were procured from two different locations namely Bikaner and Barmer, in the Rajasthan state of India. Locally available river sand was used as an additive material. The engineering properties of both bentonites and sand were determined following the Standard (IS) soil classification system. The grain size distribution of the bentonites along with sand is presented in Fig. 1. The geotechnical properties of both the bentonite and sand are presented in Table 1. The divalent bentonite possesses high shrinkage limit and high specific surface area. The monovalent bentonite exhibited higher liquid limit, higher amount of montmorillonite content and higher cation exchange capacity. According to the IS soil classification system, the river sand was classified as a poorly graded medium sand and the bentonite were highly compressible clays. Figure 2 presents the field emission scanning electron microscope images (FESEM) of both the bentonites along with sand.

The FESEM images of the powder divalent bentonite showed the mixed structure of large number of platy and needle-like fibrous structure combination, which evidenced the mixture of both montmorillonite and palygorskite, respectively. On the other hand, the monovalent bentonite consisting large fraction of cornflake structure indicates montmorillonite minerals. The FESEM images of the sand showed the different shapes of the particles.

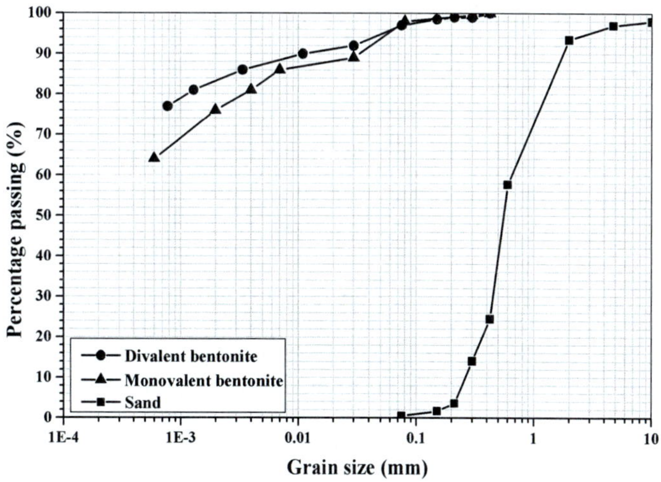

Fig. 1 The grain size distribution of the bentonites and sand

Table 1 Physico-chemical properties of the clays used in the current study

Property	Monovalent bentonite	Divalent bentonite	Sand
Specific Gravity	2.76	2.64	2.62
Liquid limit	220	172	–
Plastic limit	43	52	–
Shrinkage limit	12	26	–
Cation exchange capacity	65	48	–
Smectite content	76	42	–
Specific surface area (m^2/g)	386	520	–
Coefficient of Uniformity, C_u	–	–	2.63
Coefficient of Curvature, C_c	–	–	1.02
Soil classification	CH	CH	SP

Fig. 2 The FESEM images of both the bentonites and sand

2.2 *Specimen Preparation for Suction and Swelling Pressure Tests*

The ratio of the dry mass of sand in a bentonite–sand mixture to the total weight of the mixture is defined as the sand ratio [17].

$$R_s = \frac{m_s}{m_s + m_b}$$

Where R_s is the sand ratio, m_b and m_s are the dry mass of bentonite and sand in the mixture, respectively. Different sand ratios of 40, 50, and 60% bentonite–sand mixtures were prepared by taking the required quantity of bentonite and sand thoroughly mixed to obtain homogenous mixtures. In order to conduct suction tests, the bentonite–sand mixtures were mixed with different moisture contents. The saturated mixture slurries were placed in airtight plastic bags, and these bags were kept in a desiccator to control the moisture exchange from the external environment. After equilibration of 7 days, the soil specimens were taken out from the desiccator and used for suction measurement. The suction was measured using a chilled mirror dewpoint hygrometer (WP4), based on the measurement of the relative humidity of the samples. The soil specimen's cups have been designed to be fitted directly into the WP4C device drawer. Suction values are measured for the various interval of time depending on the drying condition. To obtain more reliable results two series of measurements were taken.

For conducting swelling pressure tests, the bentonite–sand mixtures were compacted at a constant dry density of 1.6 Mg/m^3 using different sand ratios, i.e., 40, 50, and 60%. In order to compare the effect of sand content on the swelling pressure of the bentonites, the swelling pressure tests were also conducted on pure bentonites without adding any sand fraction. The soil specimens of size 20 mm height and diameter 60 mm were prepared by using a specially manufactured cylindrical mold of 55 mm height and 60 mm diameter. In order to achieve the targeted dry density, the pre-calculated quantity of bentonite–sand mixtures in powder form were placed inside the special mold consisting of cell ring at the bottom. The soil specimens were compacted using a static uniaxial compaction process. The compressive force was applied statically at the top portion of the plug till the flange is in contact with the barrel of the mold. The compaction pressure mainly depends on the type of bentonite and sand fraction. The applied compaction stress was noted to be decreased from 44 to 13 MPa to the corresponding increase in the sand ratio from 0 to 60% for divalent bentonite. Similarly, for monovalent bentonite, it was noted be decreased from 7.8 to 2.1 MPa to the corresponding same sand ratios. It can be noted that as an increase in sand ratio, the amount of applied compaction pressure was noted to be decreased for both the bentonites.

3 Results and Discussion

3.1 Effect of Sand Ratio on Suction Properties of the Bentonite–Sand Mixtures

The effect of sand ratio on the total suction of both the bentonites is presented in Fig. 3. The results indicate that at a given suction value, the increase in sand ratio caused to decreasing the water holding capacity of both the bentonites. It can also be noted that at a given sand ratio, the water retention capacity of the monovalent bentonite is higher than that of divalent bentonite for the suction range lower than 3 MPa. Whereas for suction values greater than 3 MPa, the water retention capacity of the divalent bentonite–sand mixtures was found to be higher than that of monovalent bentonite–sand mixtures. It can be concluded that the higher montmorillonite content present in the monovalent bentonite formed large number of quasi-crystals and hold large amount of bound water in between crystal lattice. However, the large quantity of bound water can be easily depleted by applying suction. On other hand, the divalent bentonite consisting both platy and needle-like fibrous structure combination (Fig. 2), which evidenced the mixture of both montmorillonite and palygorskite, respectively.

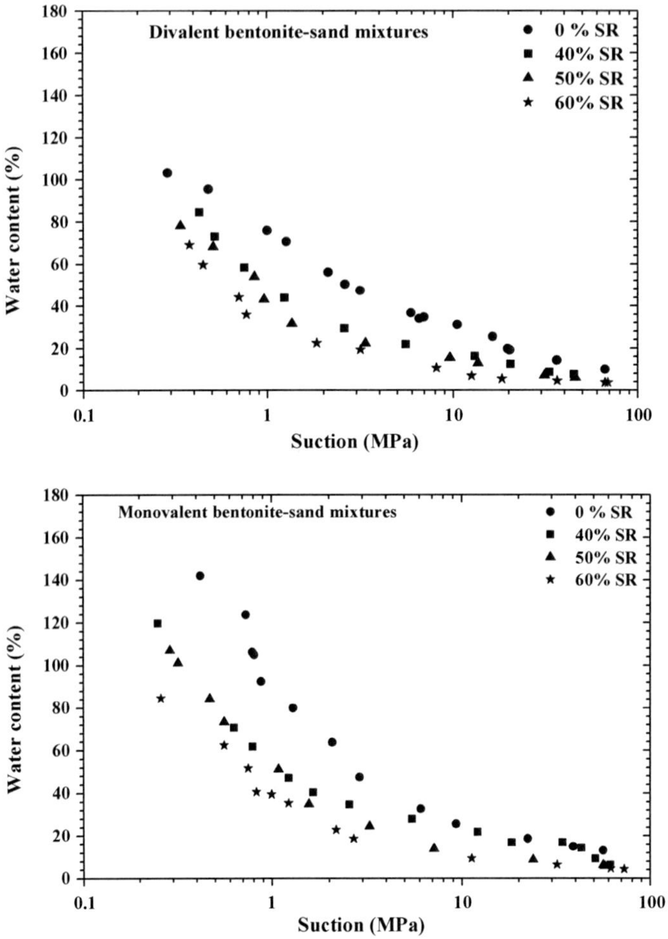

Fig. 3 The effect of sand ratio on total suction of the bentonite–sand mixtures

Therefore, the needle-like clay particles were formed limited quasi-crystals and hold a limited amount of bound water in between crystal lattice. The higher water holding capacity of the divalent bentonite at higher suction of more than 3 MPa indicates that although the divalent bentonite having limited water holding capacity at lower suctions, it was tightly held against applied suction ranges.

At higher suction between 10 to 100 MPa, the water holding capacity was noted to be almost similar for sand ratios of 40, 50, and 60%. However, a notable decrease in water holding capacity was noted for the sand ratio between 0 and 40%. On other hand, for monovalent bentonite–sand mixtures, with increase in sand ratio the water holding capacity marginally decreased for the suction range between 10 to 100 MPa. For both the bentonites, the increase in sand ratio significantly effects the water retention capacity for the lower suction range below 3 MPa.

3.2 *Effect of Sand Ratio on Swelling Pressure Development of the Bentonite–Sand Mixtures*

Figure 4 showed the elapsed time versus swelling behavior of both the bentonite–sand mixtures at various sand ratios of 0, 40, 50 and 60%. From Fig. 4 it can be observed that at sand ratio of 0%, the swelling pressure of the divalent bentonite increased rapidly and attained equilibrium within 24 h. On other hand, the swelling pressure of monovalent bentonite was noted to progressively increase at the initial stage of the test, followed by decrease in swelling pressure was noted at the intermediate stage and finally increased again. From the Fig. 4 it can be observed that as an increase in sand ratio, the collapse behavior at the intermediate stage was noted to decrease

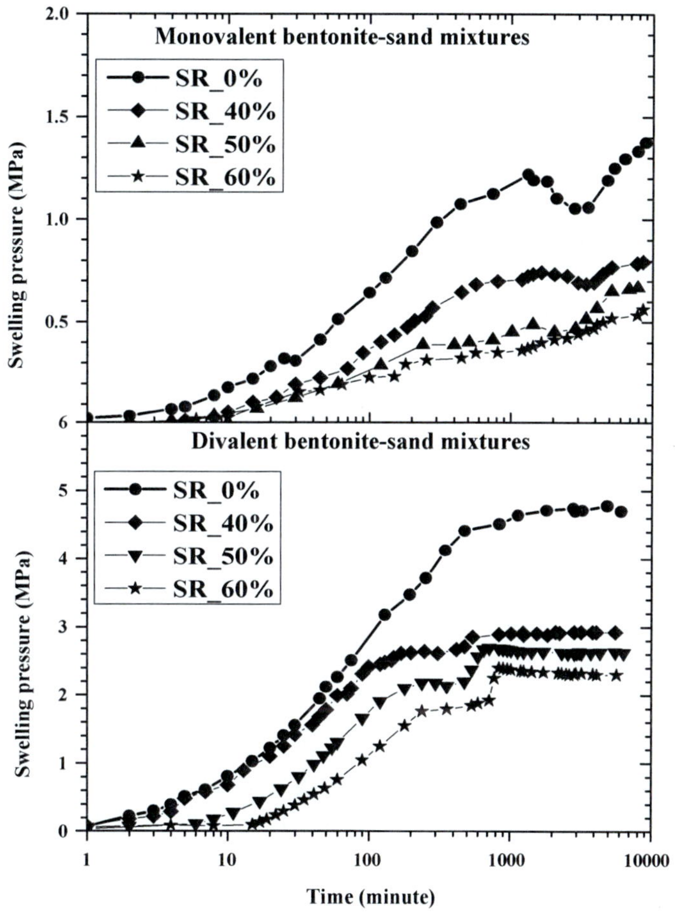

Fig. 4 Time-swelling pressure behavior of both the bentonite–sand mixtures with varying sand ratio

for monovalent bentonite. On other hand, the addition of sand lead to a marginal collapse behavior during swelling pressure development of divalent bentonite.

The difference in swelling pressure behavior can be explained that, at a given sand ratio, the higher surface area of divalent bentonite is attributed to applied higher compaction pressure on soil specimen preparation for swelling pressure test. In comparison, the applied compaction stress was noted to be lower for monovalent bentonite–sand specimens. The higher compaction pressure induced to decrease the macropore volume of the compacted bentonites specimens [8, 9]. However, at a given dry density, the increase in sand fraction induced to increasing the formation of the macropore quantity [16]. Therefore, for divalent bentonite at the higher sand ratio, the increase in sand fraction induced to increase in the macropore fraction resulted in a marginal collapse in swelling pressure.

On other hand, the lower compaction stress was applied on monovalent bentonite even at sand ratio of 0%. Further, as increase in sand ratio, the applied compaction pressure decreased further. Therefore, the lower applied compaction stress were induced the existence of a larger fraction of big macropores in monovalent bentonite–sand mixtures. Increase in sand fraction leads to an increase in the macropores fraction further. Therefore during hydration, the swelling of monovalent bentonite fills the macropores and developed lower swelling pressure as compared to divalent bentonite. Further, the addition of sand the swelling pressure also decreased because the lower amount of bentonite quantity is insufficient in filling the big macropores developed by the sand matrix [14]. Therefore, the lack of filling of pores induced to decrease the collapse behavior in monovalent bentonite–sand mixtures.

3.3 *Effect of Sand Ratio on Swelling Pressure of the Bentonite–Sand Mixtures*

The effect of sand ratio on the swelling pressure of the bentonite–sand mixtures is presented in Fig. 5.

It can be noted that swelling pressure decreases with an increase in the sand ratio. The swelling pressure was noted to be 4.71, 2.92, 2.62, and 2.31 MPa to the corresponding sand ratio of 0, 40, 50, and 60%, respectively, for divalent bentonite. Similarly, for monovalent bentonite, the swelling pressure was noted to be 1.37, 0.79, 0.65, and 0.56 MPa respectively for the same sand ratio. From the test results, it can be concluded that the effect of sand ratio on swelling pressure of the monovalent bentonite is more significant than that of divalent bentonite. For both the bentonites, the swelling pressure was noted to decrease exponentially with an increase in sand ratio.

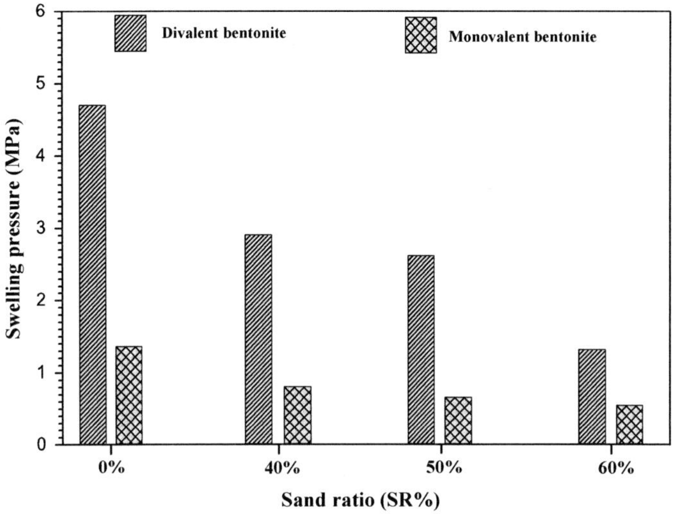

Fig. 5 The effect of sand ratio on swelling pressure of the bentonite–sand mixtures

4 Conclusions

This paper presents the effect of sand ratio on suction and swelling pressure of two bentonite–sand mixtures. The experimental results showed that the water holding capacity of the monovalent bentonite–sand mixtures is higher at lower suction range of 3 MPa, whereas at higher suction range, the divalent bentonite–sand mixtures having greater water retention capacity. At a given suction, the increase in sand ratio leads to decreasing the water retention capacity of both the bentonite–sand mixtures. At a given sand ratio, the swelling pressure exhibited by divalent bentonite–sand mixture was significantly higher as compared to monovalent-sand mixtures. For both the bentonites the swelling pressure was noted to be decreased with an increase in sand ratio.

References

1. Agus SS, Arifin YF, Tripathy S, Schanz T (2013) Swelling pressure–suction relationship of heavily compacted bentonite–sand mixtures. Acta Geotech 8:155–165
2. Bag R, Jadda K, Srikanth RN (2018) Effect of sand ratio on swelling pressure and hydraulic conductivity of an Indian bentonite-sand mixture. Unsat 2018:1–6
3. Chen Z-G, Tang C-S, Zhu C, Shi B, Liu Y-M (2017) Compression, swelling and rebound behavior of GMZ bentonite/additive mixture under coupled hydro-mechanical condition. Eng Geol 221:50–60
4. Cui SL, Zhang HY, Zhang M (2012) Swelling characteristics of compacted GMZ benton-ite-sand mixtures as a buffer/backfill material in China. Eng Geol 141–142:65–73

5. Dixon DA, Gray MN, Thomas AW (1985) A study of the compaction properties of potential clay–sand mixtures for use in nuclear fuel waste. Eng Geol 21:247–255
6. Fan R, Du YJ, Reddy K, Liu SY, Yang YL (2014) Compressibility and hydraulic conductivity of clayey soil mixed with calcium bentonite for slurry wall backfill: initial assessment. Appl Clay Sci 101:119–127
7. Gatabin C, Talandier J, Collin F, Charlier R, Dieudonné A-C (2016) Competing effects of volume change and water uptake on the water retention behaviour of a compacted MX-80 bentonite/sand mixture. Appl Clay Sci 121–122:57–62
8. Jadda K, Bag R (2020) Effect of initial compaction pressure and elevated temperature on swelling pressure of two Indian bentonites. Environ Earth Sci 79:197
9. Jadda K, Bag R (2020) Variation of swelling pressure, consolidation characteristics and hydraulic conductivity of two Indian bentonites due to electrolyte concentration. Eng Geol 272:105637
10. Komine H (2010) Predicting hydraulic conductivity of sand bentonite mixture backfill before and after swelling deformation for underground disposal of radioactive wastes. Eng Geol 114:123–134
11. Ravi K, Rao SM (2017) Estimation of Solute Concentrations in Micro-pore and Macro-pore Solutions of Compacted Bentonite-Sand Mixtures. Geotech Geol Eng 35:517–525
12. Ravi K, Rao SM (2013) Influence of infiltration of sodium chloride solutions on SWCC of compacted bentonite-sand specimens. Geotech Geol Eng 31:1291–1303
13. Sivapullaiah PV, Sridharan A, Stalin VK (1996) Swelling Behavior of soil-bentonite mixtures. Can Geotech J 33(5):808–814
14. Srikanth V, Mishra AK (2016) A laboratory study on the geotechnical characteristics of sand-bentonite mixtures and the role of particle size of sand. Int J Geosynth Gr Eng 2:3
15. Sun D, Zhang J, Zhang J, Zhang L (2013) Swelling characteristics of GMZ bentonite and its mixtures with sand. Appl Clay Sci 83–84:224–230
16. Wang Q, Tang AM, Cui YJ, Delage P, Barnichon JD, Ye WM (2013) The effects of technological voids on the hydro-mechanical behaviour of compacted bentonite-sand mixture. Soils Found 53:232–245
17. Xu L, Ye WM, Chen B, Chen YG, Cui YJ (2016) Experimental investigations on thermo-hydro-mechanical properties of compacted GMZ01 bentonite-sand mixture using as buffer materials. Eng Geol 213:46–54
18. Zhang M, Zhang HY, Cui SY, Jia L, Chen H (2012) Engineering properties of GMZ bentonite–sand as buffer/backfilling material for high leval waste disposal. Eur J Environ Civil Eng 16(10):1216–1237

A Study on the Application of Lightweight Deflectometer During the Construction of Low Volume Road in India

Vinod Kumar Adigopula, Chandra Bogireddy, and Rakesh Kumar

Abstract Most of the road construction projects in developing countries like India are still adopting density base quality control approach. Density base method is time-consuming and material properties like stiffness or moduli could not monitor through conventional density base method. In situ stiffness measuring device triggers to use lightweight deflectometer (LWD) method to evaluate the compaction characteristics of the unbound layer. In this study, one experimental stretch was constructed in the state of Gujarat to reveal the feasibility of application of LWD in the Indian context. The experimental stretch was low volume road and compaction process was monitored on subgrade, base and surface layers. Total of 138 test points were performed on considered layers. In situ fixing of stiffness value the methodology was elaborated and target stiffness values were determined for unbound layers. LWD target stiffness value of subgrade and subbase was found to be 30 and 47 MPa respectively. LWD can be used as in situ stiffness measuring device during the construction of low volume roads and conclusion was made based upon the coefficient of variance (COV). COV of all the layers were found to be in the range of 1.03% to 17.48% respectively. Frequency of LWD test locations can be increased and it will improve the overall quality of compacted layers.

Keywords Quality control · Low volume road · Lightweight deflectometer

---

V. K. Adigopula (✉)
Sr. Assistant Professor, Department of Civil Engineering, Madanapalle Institute of Technology and Sciences, Angallu, Chittoor District 517325, Andhra Pradesh, India
e-mail: vinodkumara@mits.ac.in

C. Bogireddy
Assistant Professor, Department of Civil Engineering, Vardhaman College of Engineering, Shamshabad, Hyderabad 501218, Telangana, India

R. Kumar
Professor, Department of Civil Engineering, Sardar Vallabhbhai National Institute of Technology, Surat 395007, Gujarat, India

A. Garg et al. (eds.), *Proceedings of the 1st Indo-China Research Series in Geotechnical and Geoenvironmental Engineering*, Lecture Notes in Civil Engineering 123,
https://doi.org/10.1007/978-981-33-4324-5_8

1 Introduction

Most accessible road network in India is categorized as low volume road (LVR) with a percentage share of 80% [1]. These roads are design to carry less traffic volumes (say less than or equal to 2 million standard axles) [2]. Under the vison of Prime minster of India rural connectivity program (PMGSY) launched in the year 2000 with an objective of connecting all rural habitation areas and vast length of rural roads was constructed. LVR are usually constructed with local availability material and mostly constructed with granular layers with or without a thin asphalt surface. During the design process it does not include any mechanistic empirical approach. Some of the studies reveal that rural transport sector is poverty-stricken of funds, in absence of practical specifications and guide lines for design, construction and maintenance of LVR not been provide satisfactory services [3]. In the report of NCHRP 2004, it is releveled that adoption of mechanistic empirical pavement design approach there is a need for developing and using alternative compaction quality control/quality assurance (QC/QA) procedures to provide stiffness or strength base measurements rather than conventional density base method [4]. Many strength or stiffness measuring devices are being used by agencies for monitoring QC/QA in road construction practices such as, plate load test (PLT), falling weight deflectometer (FWD), dynamic cone penetrometer (DCP) and lightweight deflectometer (LWD). Similarly, in situ density, moisture content and percentage of relative compaction was monitored by using sophisticated instruments like nuclear density gauge (NDG). In India, we still adopt conventional core cutter method or sand replacement method for monitoring in situ compaction of unbound layers.

The study described herein focused in particular on the use of two different devices for compaction monitoring on different layers considered in this study. Among the one is stiffness measuring device LWD and another one is in situ density and moisture content measuring device NDG. In India very limited study was focused on application of LWD and NDG during construction [5]. Study focused on to develop an understanding of the behavior of LWD and NDG in situ test for compaction control and fixing target stiffness values on natural subgrade and base layer. An experimental LVR research study was conducted in the state of Gujarat in the summer of 2017. During construction of LVR the compaction process was monitored using LWD and NDG.

1.1 *Objective*

The objective of this paper presents in situ test measurements from the study that was performed using LWD and NDG. The stiffness values for different layers was presented and finally fixing target stiffness values on unbound layers were discussed.

2 Construction of Study Area

Experimental study described in this paper was performed at Padra village. A 320-m-long × 3.75-m-wide was constructed using selected fill materials like natural subgrade, granular subbase, Water Bound Macadam (WBM) and thin asphalt layer. The machineries used during construction was JCB backhoe loader was used to fill the respective material, bulldozer and vibratory compactor model ESCORT EC5250 was used as a compactor. Leveling method using theodolite was used to verify the compaction level of each lifts. After achieving target thickness of subgrade layer, water tank was driven through study area as required and it is to adjust the moisture content of the fill material to achieve required compaction. After achieving target subgrade thickness, granular sub base thickness (GSB) was laid for entire experimental stretch in a single day. Similarly, base layer water bound macadam grade III was compacted with the help of smooth wheel roller. Total thickness of WBM layer is 225 mm and compaction process of each layers was monitored as per MoRD specifications [6]. Unbound layers compaction passes were compacted up to 98% of relative compaction. For all the layers calibrated NDG was used to determine the in situ moisture content, density and relative compaction. Eventually, a thin asphalt layer bituminous macadam was laid over base layer.

2.1 In Situ Testing of Experimental Stretch Using LWD and NDG

During LVR construction stiffness and density based tests were performed with the help of LWD and calibrated NDG. The nuclear density gauge was calibrated as per the ASTM guidelines ASTM D7759M before measuring the in situ density [7]. The NDG model used during construction was CPN MC-3 (An Istro Tek, Las Vegas, Naveda) as shown in (see Fig. 1); test protocol as per ASTM D6938, 2015 was adopted to perform the test [8]. In situ stiffness measuring device LWD model number 3031 was selected in this research. It is simple to operate, portable and lightweight device. Stiffness values were measured as per the ASTM E2583 [9]. The LWD model 3031 used in this study was shown in (see Fig. 2).

Recent study in India, LWD test with drop weight of 10, 15 and 20 kg were used on modified subgrade layer during construction [5]. Similarly, in one study, LWD with 10 kg drop weight was used on top of base and surface asphalt layer [10]. In this study LWD with drop weight of 10 kg and 300 mm diameter of circular plate was selected. Test values were recorded using personal data assistant (PDA), it is Bluetooth connected with instrument. Composite stiffness value was predictable on the basis of the Boussinesq theory relating to deflection profile of elastic half space subjected to axisymmetric surface loading as shown in Eq. 1. While performing LWD test initial three drops was considered as a seating drops and remaining three drops are considered in analysis as an average composite stiffness value.

Fig. 1 NDG at site location

Fig. 2 LWD 3031 with 10 kg drop weight

$$E_0 = \frac{f(1 - v^2)a \times \sigma_0}{d_0} \tag{1}$$

where: E_0 = composite stiffness value, $f = \pi/2$ and 2 for rigid and flexible plate, v = poisons ratio, σ_0 = peak applied stress (kPa), a = radius of loading plate (mm), d_0 = center deflection (μm).

3 Materials

Subgrade material selected in this study was classified as a CI (clay of medium plastic) type of soil. A total of 16 soil samples were collected at an interval of 20 m along the experimental stretch and the corresponding soil properties results are shown in Table 1. The lead distance between borrow pit area and experimental stretch was within distance of 100 m. Test protocol conducted on unbound materials were as per Indian standards [11–14].

Table 1 Subgrade soil properties

CI (Clay of medium plastic) soil properties		
S. No	Properties	Value
1	Specific gravity	2.61
	Consistency limits	
	LL (%)	41
	PL (%)	24
	PI (%)	17
2	Compaction properties (standard compaction)	
	MDD (kg/m^3)	1825
	OMC (%)	14.77
3	CBR value (%)	
	At optimum moisture content	14.80
	4 days soak condition	3.82
4	Grain size distribution (%)	
	Gravel	6.61
	Sand	13.21
	Silt and clay	80.18
5	Free swell index (%)	32.18
6	Unconfined compressive strength (kPa)	325

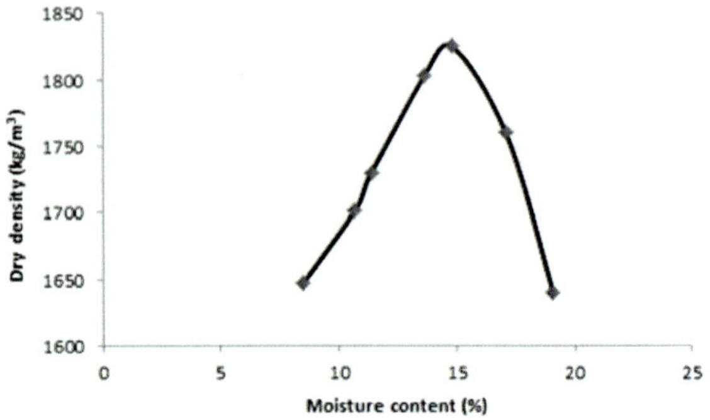

Fig. 3 Compaction curve of subgrade soil

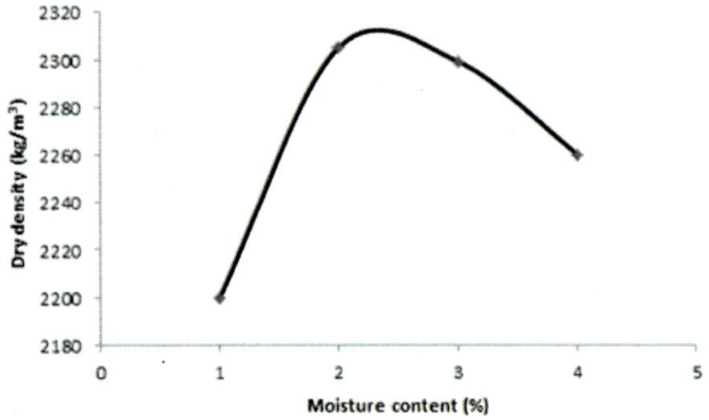

Fig. 4 Compaction curve of WBM, G-III

Subgrade is compacted in 2 lifts, each lift contains 150 mm of design thickness. Base layer (Water Bound Macadam, Grade III) consists of aggregates, screening material and granular material mixed with water, and it is laid over a granular sub base layer. LWD and NDG tests were not performed on granular subbase layer due to more irregular surface area and it was difficult to perform on it. As per Ministry of rural development (MoRD) guidelines, the lift thickness of base layer is 75 mm and it is compacted in total 3 lifts. Figures 3 and 4 shows the laboratory compaction curve of natural subgrade soil and WBM grade III. WBM gradation of design blend was shown in Fig. 5a, b.

Bituminous macadam is a well graded mix consisting of coarse aggregate, filler and bitumen binder prepared in hot mix plant and laid on base layer using mechanized paver. Bituminous macadam layer was compacted in 50 mm thick lift and its grain size distribution is shown in Fig. 5c.

4 Fixing in Situ Target Stiffness and Density Value

Target stiffness values are an important prerequisite value that has to determine at site based on material properties like in situ moisture content and density values. The German Federal Ministry of Transport suggest target stiffness values for different

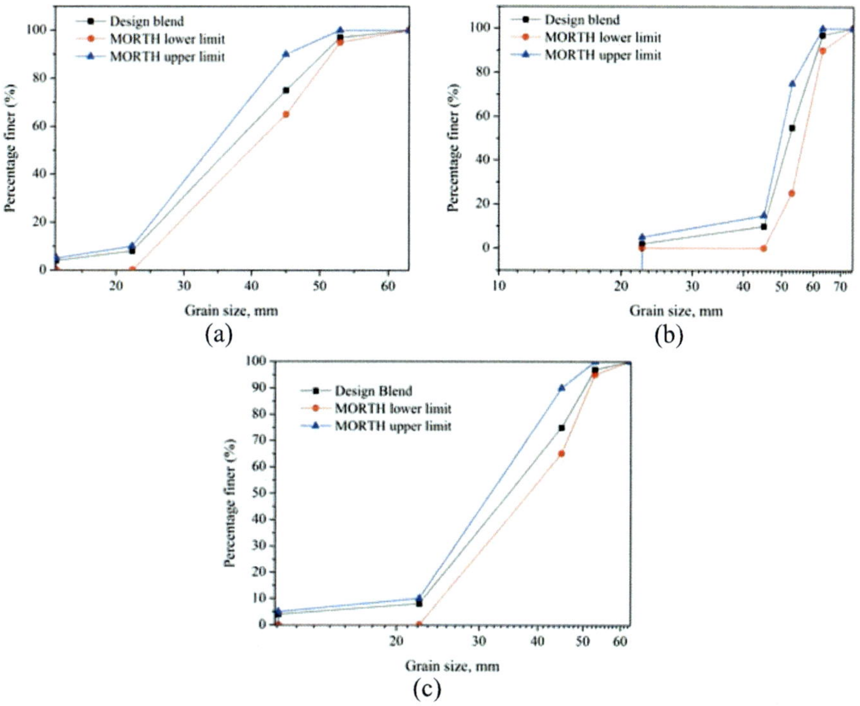

Fig. 5 Grain size distribution: **a** coarse aggregate for WBM (G-III), **b** screening of WBM, **c** bituminous macadam layer

types of soil group [15]. Target stiffness value of 80 MPa was suggested by UK highway agencies for 24 h of construction, prior to laying the bitumen surface [16]. Schwartz [17] performed LWD test on proctor mold at different moisture content to determine target stiffness value. In this experimental stretch 320 m length and 3.75 m wide was bifurcated into two sections. Section one deals with fixing target stiffness value, moisture and density values, similarly, section II deals with performance of evaluated section defined for assessing the target density and stiffness values as shown in Fig. 6.

In this study in situ fixing of target stiffness value and construction methodology was discussed as shown in Fig. 7. An attempt has been made by the authors in this study to evaluate the stiffness values on thin bitumen layer (50 mm) for the sake of brevity of results. Predetermined laboratory calculated maximum dry density (MDD) and moisture content values using Proctor test under controlled conditions as per the Indian standard guidelines. On the basis of laboratory results MDD, OMC, the authors monitored the compaction process specifically on the basis of energy level of compaction until required the laboratory values obtained in section I. Target in situ density and stiffness values were considered based on repeated test performed on target construction section I.

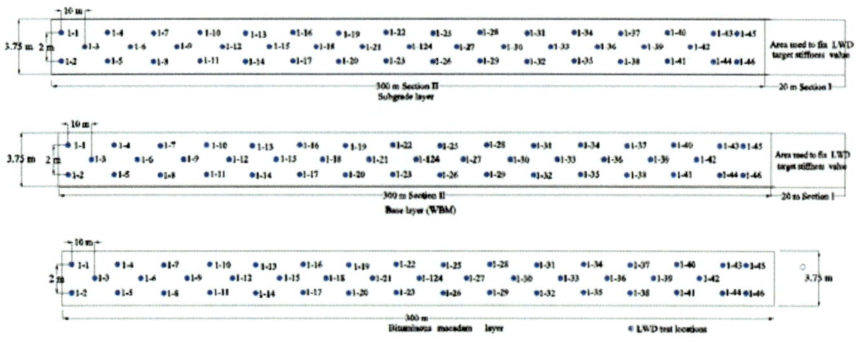

Fig. 6 Schematic view of LWD test plan on experimental stretch

In situ density and stiffness values achieved on target section I was performed as per respective protocol standards. Average target values obtained with respect to energy level of compactions at section I is shown in Table 2.

Table 3 results shown are descriptive statistics of the average stiffness and NDG density of unbound layers for evaluated section. For easy identification of target stiffness values calibration chart was developed between the LWD stiffness value and percentage of relative compaction (RC). Developed calibration chart for subgrade and base layer are shown in Fig. 8. As per road construction guidelines MoRD, at least 98% relative compaction should be achieved for unbound layers. So in this study 98% RC was considered. This target stiffness value can be used as a quality control check of compacted subgrade and base layers.

5 Assessment of Mean Target Density and Stiffness Values

The mean target NDG in situ density and stiffness values determined for subgrade and WBM layers were measured on the evaluation stretch II. The test locations were selected on the experimental stretch at an interval of 10 m to measure stiffness values on subgrade and WBM layers. Eventually, NDG test was performed at each test location to evaluate the target NDG in situ density and moisture content values along the evaluation stretch II. In addition, to density moisture content also plays a critical role in compaction. It is important to maintain the in situ moisture content of the experimental section because moisture content of the materials will have a significant effect on stiffness value [18–20]. Similar to the LWD test for each layer in test pad section I, the test locations selected for an experimental stretch II was evaluated with similar drop weight and height to retain the consistency of the test results. Figure 9 shows the achieved.

LWD stiffness values measured on the subgrade, WBM base and surface layers on section II. NDG test was conducted in a longitudinal direction of 2 m away from the LWD test location in both the sections. Descriptive statistics of LWD and NDG

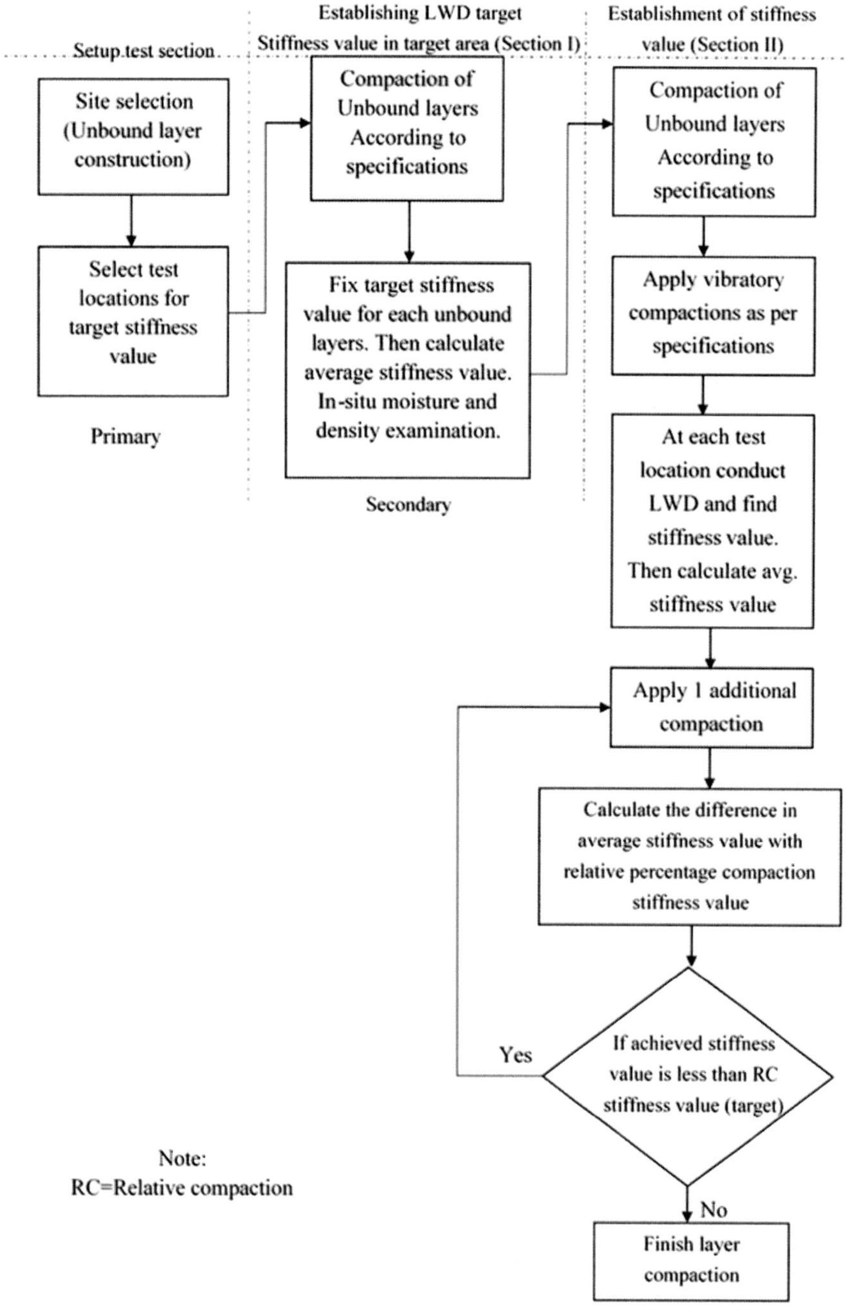

Fig. 7 Fixing target stiffness value and construction methodology

Table 2 Target in situ density and stiffness values of typical test location section I

Energy levels for compaction	Natural subgrade			Water bound macadam layer		
	γ_{d_NDG} (kg/m^3)	RC[a] (%)	$E_{0\_(LWD)}$ (MPa)	γ_{d_NDG} (kg/m3)	RC[a] (%)	$E_{0\ (LWD)}$ (MPa)
Grader I pass	1751	95.82	18	2210	95.67	42
Grader II pass	1762	96.31	21	2240	96.96	43
Grader III pass	1769	96.70	23	2252	97.48	44
Medium vibratory pass	1793	98.06	30	2266	98.09	47
High vibratory pass	1829	100	37	2330	100.86	51

Note Maximum dry density for WBM = 2310 kg/m^3; optimum moisture content = 4.4%; maximum dry density for natural subgrade = 1830 kg/m^3; optimum moisture content = 14.8%; [a]RC (Relative compaction) as compared with maximum dry density

Table 3 Mean target in situ density and stiffness values for evaluation experiment stretch section-II

Descriptive statistics	Natural subgrade		Water bound macadam layer	
	γ_{d_NDG} (kg/m^3)	E_0 (MPa)	γ_{d_NDG} (kg/m^3)	E_0 (MPa)
No. of test points	10	10	10	10
Mean	1796	30	2262	45
Median	1790	24	2254	41
Maximum	1830	41	2312	66
Minimum	1756	23	2012	28
Standard deviation	40.36	4.29	38.98	6.24
Coefficient of variance (%)	2.247	14.2	1.723	13.86

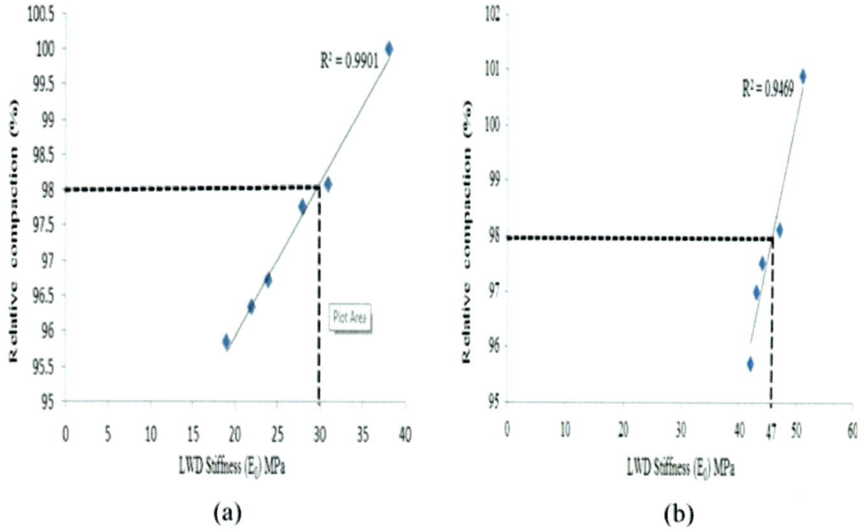

Fig. 8 Calibration chart showing target stiffness values: **a** subgrade layer, **b** base layer (WBM)

test data were shown in Table 4. After careful monitoring the compaction process, achieved stiffness values are shown in Fig. 9. As shown in Fig. 9 significant point to point variation was witnessed for measuring stiffness values on respective layer materials. It is very difficult to achieve exact target stiffness value on respective compacted materials. But coefficient of variance should not be more deviate for reliable test. Based on Table 5 LWD had the greatest tendency to measure the stiffness of unbound layer that enhance 'significant emphases' in the results that coefficient of variance less than 17.48%. At 98% relative compaction at evaluation section the NDG moisture content that falls +1% that of laboratory determined optimum moisture content on subgrade layer. Similarly, NDG moisture content on base layer were obtained ±2% on evaluated section.

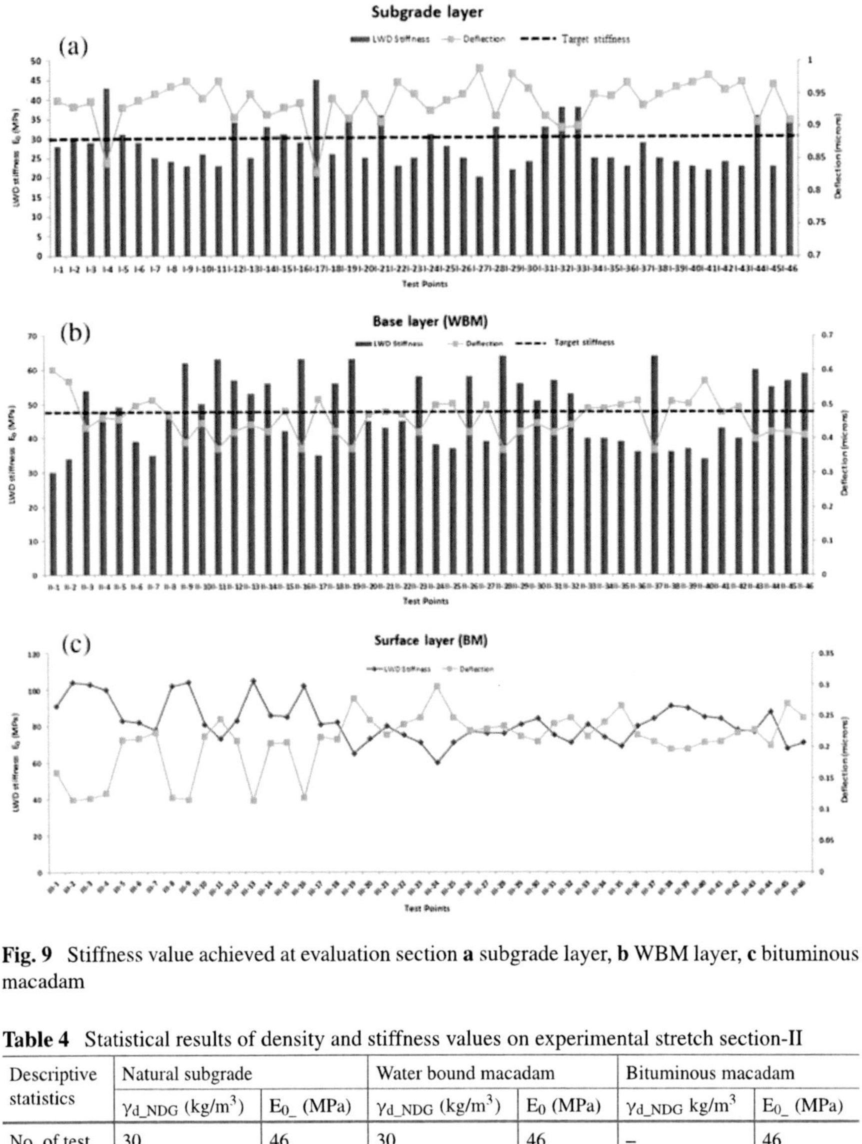

Fig. 9 Stiffness value achieved at evaluation section **a** subgrade layer, **b** WBM layer, **c** bituminous macadam

Table 4 Statistical results of density and stiffness values on experimental stretch section-II

Descriptive statistics	Natural subgrade		Water bound macadam		Bituminous macadam	
	γ_{d_NDG} (kg/m^3)	$E_{0\_}$ (MPa)	γ_{d_NDG} (kg/m^3)	E_0 (MPa)	γ_{d_NDG} kg/m^3	$E_{0\_}$ (MPa)
No. of test points	30	46	30	46	–	46
Mean	1790	28.41	2260	48.239	–	86.04
Median	1785	26	2257	48.5	–	85
Standard deviation	35.32	5.87	45.36	10.172	–	11.038
Minimum	1730	20	2013	30	–	60
Maximum	1832	45.00	2316	64	–	105

Table 5 Summary of test results from all layers

Pavement layers	Number of tests	COV[a] E_{0_LWD} (%)	
		Minimum	Maximum
Natural subgrade	46	1.03	9.9
Water bound macadam	46	2.56	16.26
Bituminous macadam	46	1.21	17.48

Note [a]*COV* Coefficient of variation

Meehan et al. [20] revealed that from this experimental study between LWD stiffness and NDG results and concluded that stiffness base test tends to show no significant difference in stiffness value with additional compaction passes (from the sixth pass of compaction obtained stiffness values are less than or equal to second pass of compaction). Reason for this scenario may be the difference between effect of moisture content and matrix suction, in real time by the filed engineers has to make a decision whether to pass or fail a given lift.

6 Results and Discussion

For any road construction projects, this methodology can be applied for setting target stiffness modulus values for the base or subgrade layers that must be attained during the different construction stages. With such investigations, a relation can be established between in situ density and stiffness values of unbound layers for various passes of the vibratory compactor. As shown in Fig. 8 a moderate strong relationship shown between density and stiffness values at certain compaction passes. In Fig. 9a, b it is believed to be caused by changes in partially saturated soils, as soil suction that occurs as the slight soil moisture content changes, which changes effective stress between the soil particles and affects the associated deflection response of the soil under load. Recent study claimed that LWD target stiffness values obtained from the in situ test data for subgrade and base layers are 30 and 50 MPa respectively [21]. Eventually, the stiffness values obtained in this investigations are closely matched and this strengthen the statements made by other researchers [5, 21–23]. The COV of stiffness values for both the sections were found to be less than 18% (see Table 2 and 4), such COV is acceptable for performing in situ stiffness base measurements in QC using this device. Based on conclusions upon previous studies and this investigation, maybe it is recommendable to determine the LWD target stiffness value by considering in situ conditions for the sake of brevity and consistency results. Authors are recommended that LWD cannot be used for thick asphalt layers. In this study LWD test performed on bituminous layer only for the sake of knowledge and no detail discussion will be highlighted.

7 Conclusion

Lightweight deflectometer and NDG test was performed on subgrade, base and surface layers to assess the feasibility of stiffness evaluating during the construction of unbound layers. The testing program includes LWD tests on 46 test points of subgrade layer, base layer and bituminous macadam layer respectively. Based on the experimental results the following conclusions were drawn in this study.

1. Loos lift of unbound layers were compacted gradually with smooth wheel vibrator and it is found that stiffness and density values were increased linearly to an relative compaction 95–100%. Considering the boundaries of relative compaction 98% ensure proper compaction and prepared at the optimum moisture content of unbound layers as suggested in MoRTH guidelines.
2. For each unbound layer that was tested, the LWD measured stiffness value had a less coefficient of variance (i.e., 1.03–9.9% for subgrade layer and 2.56–16.26% for base layer). Such a low coefficient of variance is acceptable for the compaction purpose during quality control.
3. Significant point to point variation was noticed on each layer. This variation exhibits even when there is no significant change in material that have been used in the experimental stretch.
4. Stiffness based quality control assessment using LWD is found to be simple by preparing a representative experimental stretch by defining the target stiffness values corresponding to the in situ density and moisture content values for the unbound layers. This further facilitates the performance predictions of the pavement layers as well.
5. The overall LWD stiffness modulus of subgrade and base layer compacted to a fixed relative compaction 98% by the high vibratory compaction of 5th pass was 20–45 MPa and 30–64 MPa respectively. These test results are highly dependent on the material and instrument that was used. Therefore, LWD is feasible to conduct during low volume road construction.
6. Figure 8 showed a good correlation between LWD stiffness value and density obtained from nuclear density gauge. A similar calibration graph can be used by the field engineers to perform quality control of unbound layers at construction site location.

References

1. MORTH (Ministry of Road, Transport and Highways) (2012) Basic road statistics of India. Transportation Research Wing, New Delhi, India
2. IRC (Indian Road Congress) (2015) Guidelines for the design of flexible pavements for low volume roads. IRC: SP 72, New Delhi, India
3. SahooUC, Sudhakar Reddy K (2011) Performance criterion for thin-surface low-volume roads.Transp Res Rec 2203(1):178–185

4. NCHRP (2004) Guide for mechanistic-empirical design of new and rehabilitated pavement structures. Final report for project 1-37A. Part 1 & Part 3. Chap. 4. National Cooperative Highway Research Program, Transportation Research Board, National Research Council, Washington, D.C.
5. Kumar R, Adigopula Jr VK, Guzzarlapudi SD (2017) Stiffness-based quality control evaluation of modified subgrade soil using lightweight deflectometer. J Mater CivEng29(9):04017137
6. Ministry of Rural Development (MoRD) (2014) Specifications for rural roads, 5th edn. Indian Road Congress, New Delhi, India
7. ASTM (2014) Standard guide for nuclear surface moisture and density gauge calibration. ASTM D7759M, West Conshohocken, PA
8. ASTM (2015) Standard test method for in-place density and water content of soil and soil-aggregate by nuclear methods. ASTM D6938, West Conshohocken, PA
9. ASTM (2007) Standard test method for measuring deflections with a lightweight deflectometer (LWD). ASTM D2583, West Conshohocken, PA
10. Umashankar B, Hariprasad C, Kumar TG (2015) Compaction quality control of pavement layers using LWD. J Mater Civ Eng. https://doi.org/10.1061/(ASCE)MT.1943-5533.0001379,04015111-1(9)
11. IS (Indian Standard) (2006) Determination of liquid and plastic limit. IS: 2720 (Part 5). Methods of tests for soils. Bureau of Indian Standards, New Delhi, India
12. IS (Indian Standard) (2001) Determination of shrinkage factors. IS: 2720 (Part 6). Methods of tests for soils. Bureau of Indian Standards, New Delhi, India
13. IS (Indian Standard) (2007) Classification and identification of soils for engineering purposes. IS: 1498-1970. Bureau of Indian Standards, New Delhi, India
14. IS (Indian Standard) (2011) Method of test for soils, determination of water content-dry density relation using light compaction. IS: 2720 (Part 7).Bureau of Indian Standards, New Delhi, India
15. German Federal Ministry of Transport (1994) Additional technical contractual conditions and guidelines for earthwork in road construction. German Federal Ministry of Transport, Koln, Germany
16. Highways Agency (2006) Design guidance for road pavement foundations (Draft HD 25). Interim advice note 73. Highways Agency, London
17. Schwartz CW, Afsharikia Z, Khosravifar S (2017) Standardizing lightweight deflectometer modulus measurements for compaction quality assurance. No. MD-17-SHA-UM-3-20. State Highway Administration, Maryland
18. Rahman F, Hossain M, Hunt MM, Romanoschi SA (2008) Soil stiffness evaluation for compaction control of cohesion less embankments. Geotech Test J 31(5):1–13
19. Hossain MS, Apeagyei AK (2010) Evaluation of the light weight deflectometer for in situ determination of pavement layers moduli. FHWA/VTRC 10-R6. Virginia Transportation Research Council, Charlottesville, VA
20. Meehan CL, Tehrani FS, Vahedifard F (2012) A comparison of density-based and modulus based in situ test measurements for compaction control. Geotech Test J 35(3):387–399
21. Nunn ME, Brown A, Weston D, Nicholls JC (1997) Design of long-life flexible pavements for heavy traffic. TRL report 250. Transport Research Laboratory, Crowthorne, Berkshire, UK
22. Highways Agency (2006) Design guidance for road pavement foundations. Draft HD 25. Interim advice note 73, London
23. Kumar A, Guzzarlapudi SD, Kumar R (2017) Structural evaluation of flexible pavements using non-destructive techniques in low volume road. In: GeoMEast 2017: advancement in the design and performance of sustainable Asphalt pavements, pp 168–184

A Note on Partial Safety Factors for In-situ Shear Strength Parameters of Rock-Mass

Shashank Pathak and G. V. Ramana

Abstract The design of retaining structures founded on jointed rock-mass must ensure their safety against sliding instability. For this purpose, generally, in-situ direct shear tests are conducted and shear strength parameters (cohesion and friction-angle) are determined using the popular Mohr–Coulomb criterion. However, due to assumptions associated with such a simple shear strength criterion and various other uncertainties, random errors and limitations associated with in-situ testing, the shear strength parameters are also associated with uncertainties. The present study discusses a practise-friendly procedure to account for such uncertainties and investigates the efficacy of the current design practice (IS 6512) in accounting for these uncertainties. The proposed procedure is explained through four case-studies of in-situ direct shear tests conducted in Phyllite and Granitic rock-masses and evaluating sliding safety of a hypothetical concrete gravity dam founded on those rock-masses. The study highlights the fact that IS 6512 may lead to an unsafe design in cases when uncertainties associated with in-situ shear test data are significant.

Keywords Partial factor of safety · In-situ shear test · Jointed rock-mass · IS 6512 · Cohesion · Friction-angle

1 Introduction

An insufficient shear strength of an interface between structure and geomaterial or within the geomaterial itself may cause sliding failure of retaining structures such as dams (e.g. International Commission on Large Dams (ICOLD) [8]; Hao and Li [7]. Therefore, checking the safety of design against sliding instability is an important component of design of dams (e.g. Zhou et al. [21]; Chen et al. [5];

S. Pathak
Precision Mechatronics Laboratory, Université Libre de Bruxelles, 1050 Brussels, Belgium
e-mail: shashankpathaks@gmail.com

G. V. Ramana (✉)
Department of Civil Engineering, NIT Warangal, Warangal 506004, India
e-mail: gvramana@nitw.ac.in

A. Garg et al. (eds.), *Proceedings of the 1st Indo-China Research Series in Geotechnical and Geoenvironmental Engineering*, Lecture Notes in Civil Engineering 123,
https://doi.org/10.1007/978-981-33-4324-5_9

Krounis et al. [11]). In most of the routine engineering problems, safety evaluation of design includes sliding stability analysis along the concrete over rock (generally at the founding level) and rock over rock (generally at a shear plane or otherwise within the rock-mass) interfaces and sliding is assumed to be resisted by frictional and cohesive forces acting against the horizontal driving forces. The friction and cohesion, together, determine the shear strength of the critical interface along which the sliding stability needs to be ensured. Usually, for the purpose of design two critical interfaces are considered: (i) the one between structure (usually concrete) and the geomaterial (soil or rock-mass) and another (ii) within the geomaterial itself (usually a shear plane within the rock-mass). Therefore, it is a key responsibility of field geotechnical engineers to carefully investigate and interpret the test data from in-situ shear test of the geomaterials. However, when structures are to be founded on a jointed rock-mass, the interpretation of in-situ shear test data becomes quite challenging due to inherent variability in the test data (e.g. Alonso et al. [2]). In addition to this, field tests are quite difficult to conduct due to complex rocky strata and risky approach to the test sites which allows only a limited number of tests. The limited test data is suitable for determination of shear strength parameters based on only Mohr–Coulomb criterion, though there are several other rigorous strength criteria (such as Hoek-Brown and Drucker-Prager criterion) are also available (e.g. Brown and Hoek [4]; Lin and Zhang [13]; Sheorey [18]; Yu et al. [20]; Eberhardt [6]; Alejano and Bobet [1]; Barton [3]). Mohr's condition assumes that failure depends only on maximum and minimum principle stresses and the shape of the failure envelope is linear at low stresses and non-linear at higher stresses (Labuz and Zang [12]), whereas, Coulomb's condition is based on a linear failure envelope. Thus, the Mohr-Coulomb criterion is based on the assumption of linearity at all stress levels and the effect of intermediate principal stress is also not considered (Labuz and Zang [12]). This criterion relates peak shear stress (τ_f) to applied normal stress (σ_n) as:

$$\tau_f = c + \sigma_n \tan \phi \tag{1}$$

where, c is cohesion and φ is friction-angle. For determination of the two shear strength parameters (Eq. 1), in-situ direct shear tests are conducted at concrete over rock and rock over rock interfaces in exploratory drifts or in open trenches in rock-mass (e.g. Singh [19]; Ramana et al. [17]; Pathak et al. [15]; Ramana et al. [16]). During such tests, horizontal shear stresses are applied on an interface under a constant normal stress and peak shear stresses are recorded corresponding to the normal stress. Usually, in-situ shear tests are conducted at different test locations within the exploratory drift or trench at different normal stress levels and the normal stress versus peak shear stress data is fitted in Eq. 1 that leads to determination of cohesion c and friction-angle φ. Since there are several inevitable aleatoric and epistemic uncertainties associated with (i) site preparation (such as unevenness of shearing interface), (ii) testing procedure (such as eccentricity of loads causing undesirable rotations), (iii) experimental observations (such as erroneous data recording or reading of dial gauge or malfunctioning of readout unit), (iv) limited data (due to difficulties in site preparation and project economy), (v) selection of representative site (due to subjective engineering judgement), and (vi) interpretation of test data (for example, area of the shearing interface). To account for such uncertainties associated with the estimation of design parameters, the use of factor of safety-based design approach has gained much popularity among the practising professionals due to its simplicity and intuitiveness. Such methods are especially useful when quality and quantity of the test data is poor (Kreuzer and Bury [10]). Therefore, current design standards such as IS 6512 [9] suggest to apply partial factors of safety (PFOS) to the two shear strength parameters to obtain the design shear strength parameters which are used further to calculate the factor of safety against (FOS) sliding failure i.e. ratio of resisting force to driving force. An FOS less than one indicates unsafe design and warrants appropriate revision of the design. It is to note that the PFOS defined by IS 6512 [9] accounts only for uncertainties in physical behaviour of dam-foundation system under prevalent environmental conditions. For example, under normal operating conditions (i.e. full reservoir, dry weather and normal uplift) partial factor of safety for friction-angle is given as 1.5, irrespective of concrete to rock or rock to rock interface, whereas, partial factor of safety for cohesion is given as 3.6 at concrete to rock interface and 4.0 at rock to rock interface under the assumption that foundations are thoroughly investigated. However, to the authors' knowledge there is no such study which evaluates the efficacy of IS 6512 [9] in accounting for the uncertainties related to the site-specific in-situ shear test data. In view of this, the present paper proposes a methodology to assess the suitability of the existing partial safety factors at different uncertainty levels associated with the in-situ shear test data.

1.1 Organization

This paper is organized in following four sections: (i) the first and current section introduces the paper, (ii) the second section discusses the theoretical details of the procedure proposed to account for testing related uncertainties, (iii) the third section is devoted to the illustration and application of the proposed procedure by considering in-situ shear test data from three different rock sites. This section also discusses how successful IS 6512 [9] is in accounting for the site-specific test data related uncertainties, and (iv) the last section highlights the key conclusions of this study.

2 Proposed Procedure

The uncertainties associated with in-situ shear tests would reflect on the peak shear stress versus applied normal stress data. Therefore, to account for statistical errors in the estimation of cohesion and friction-angle, it is assumed that these errors are normally distributed with a zero mean. Thus, it can be shown that, statistically, the model given by Eq. 1 refers to the expected mean trend of the failure envelope. It may also be noted that a confidence interval estimate of linear regression parameters (i.e., c and tan φ in present case) can represent the effect of all these errors and uncertainties in a probabilistic manner. The standard statistical procedure to build such confidence intervals can be followed from Montgomery et al. [14]. This confidence interval provides an upper and a lower bound such that the lower bound provides a conservative shear strength compared to the least-square estimate. In view of this, this study proposes a new PFOS on c and tan φ as:

$$\text{PFOS} = \frac{\text{Parameter}}{\text{Lower Bound}} \tag{2}$$

2.1 Basic Properties of Proposed PFOS

Following are the key characteristics of the proposed definition of partial factor of safety:

1. It is site-specific and would depend upon number of tests conducted.
2. It would also depend upon the choice of confidence level which is a subject of engineering judgement.
3. If testing related uncertainties are higher than confidence intervals would be wider. This would lead to a higher PFOS value.
4. If testing related uncertainties are minimized then confidence intervals would be narrower and a lower value of PFOS would be obtained.

3 Description of Proposed Procedure

3.1 In-Situ Shear Test Data

To illustrate the evaluation, application, and implication of the proposed PFOS, the in-situ direct shear test data available in two sources (i) Ramana et al. [17] and (ii) Ramana et al. [16] is adopted in this study. The first source is related to the in-situ shear strength parameter interpretation of weathered Phyllites rock-mass at Amochu, Bhutan.

The adopted data in this study consists of 10 direct shear tests on rock/rock (R/R) interface and 10 direct shear tests on concrete/rock (C/R) interface. On the other hand, the second source discusses the variation in R/R and C/R shear strength parameters of Granites among different geological conditions of Eastern Ghat Belt of Andhra Pradesh (EGB) and western part of Himalaya (WHI). In the current paper, 5 C/R test data from EGB and 5 C/R data from WHI are adopted and analysed. Various geological details of the tests sites and test procedure can be followed from these two references. In addition to this, these two references also provide an overview about

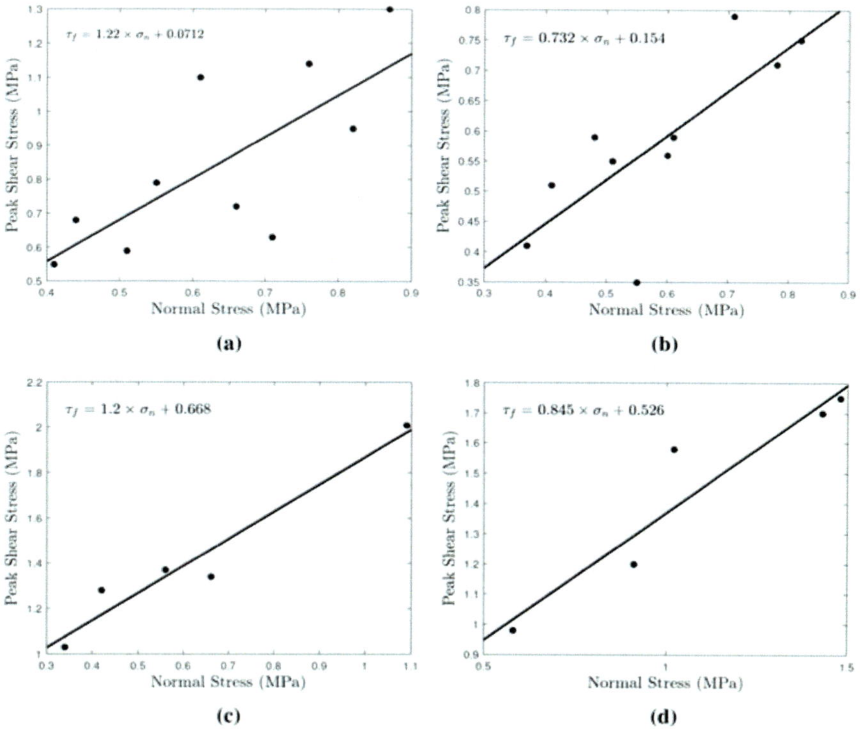

Fig. 1 Shear stress versus normal stress plot for **a** R/R and **b** C/R interfaces in weathered Phyllites rock-mass in Amouchu and **c** C/R interface in Garnetiferous-quartzo-feldspathic- biotite-gneiss rock-mass in Eastern Ghat Belts (India), and **d** C/R interface in Granite-gneiss rock-mass in Western Himalayas

the variability and uncertainties associated with in-situ shear tests which have been briefly highlighted earlier in this paper. The test data used in this study are shown in Fig. 1 along with fitted Mohr-Columb criterion.

3.2 Evaluation of Partial Factors of Safety Data

Using the proposed procedure (Eq. 2), PFOS are evaluated for cohesion (F_C) and tangent of friction-angle ((F_{Φ})) at different confidence levels (CL) for the four case-studies (Figs. 2 and 3), respectively.

Due to various uncertainties, the standard errors may be large enough in some cases which may lead to negative values of lower bound of the two shear strength parameters. In such cases, the value of PFOS is not determined as treatment of such cases is beyond the scope of present study which require that errors should be modelled by considering a positive distribution such as log-normal distribution or

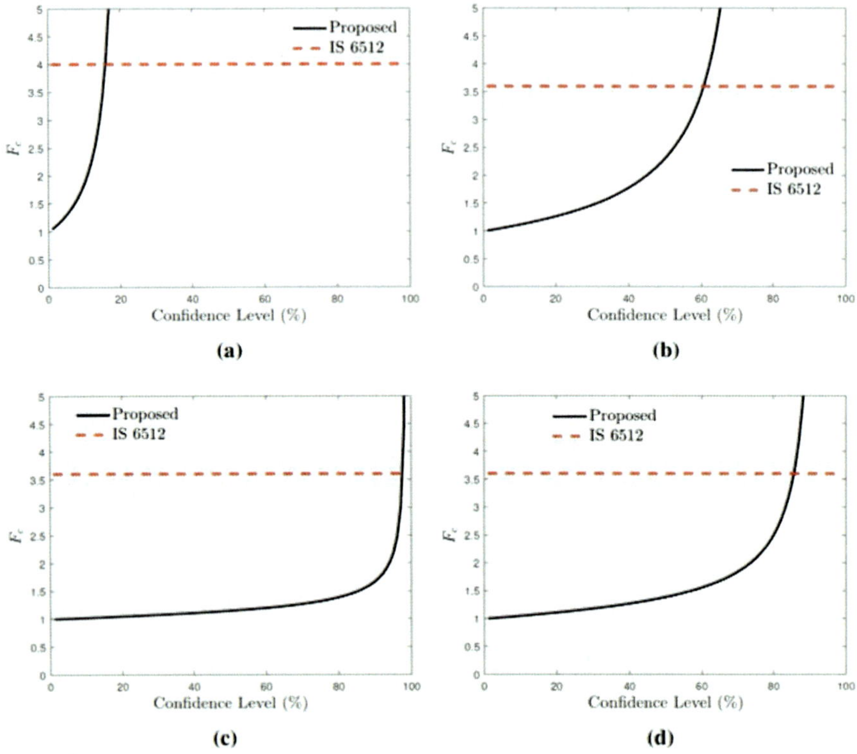

Fig. 2 Variation of PFOS of cohesion for **a** R/R and **b** C/R interfaces in weathered Phyllites rock-mass in Amouchu and **c** C/R interface in Garnetiferous-quartzo-feldspathic-biotite-gneiss rock-mass in Eastern Ghat Belts (India), and **d** C/R interface in Granite-gneiss rock-mass in Western Himalayas

truncated normal distribution. A higher CL provides a wider possibility of variation of parameter and hence higher PFOS is observed at higher CL. It can be clearly seen that the partial factors of safety suggested by IS 6512 [9] can cover the requirement of proposed PFOS upto a certain confidence level only and beyond that IS 6512 [9] may not be sufficient to cover the uncertainties associated with in-situ shear test data.

3.3 *Evaluation of Partial Factors of Safety Data*

The factor of safety Fs against sliding instability of gravity dams can be evaluated as (IS 6512 9]):

$$F = \frac{\frac{(w-u)\tan\varphi}{F_\phi} + \frac{CA}{F_C}}{\mathrm{P}} \tag{3}$$

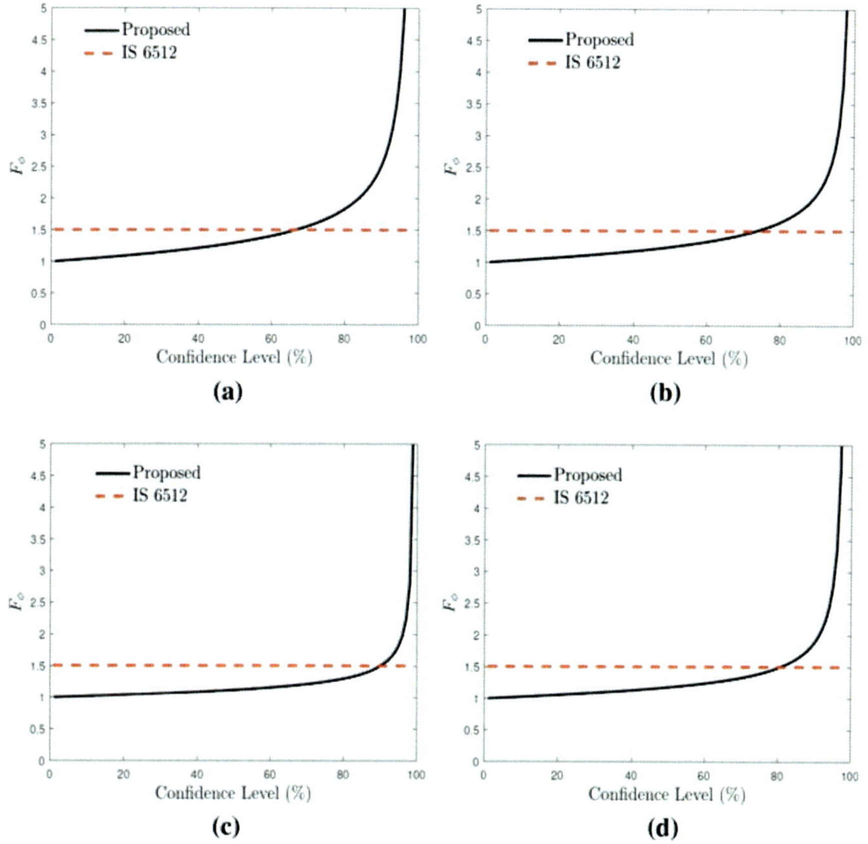

Fig. 3 Variation of PFOS of friction-angle for **a** R/R and **b** C/R interfaces in weathered Phyllites rock-mass in Amouchu and **c** C/R interface in Garnetiferous-quartzo-feldspathic- biotite-gneiss rock-mass in Eastern Ghat Belts (India), and **d** C/R interface in Granite-gneiss rock-mass in Western Himalayas

where, w = total weight of the dam, u = total uplift force, A = area under consideration for cohesion, and F_h = total horizontal force. For the purpose of illustration, let us consider an example of concrete gravity dam with a typical base width of 70 m, height of 100 m, top width of 15 m and concrete density of 25 kN/m^3.

All computations are carried out per unit length of the dam and total uplift force has been ignored. Total horizontal force has been taken only due to horizontal water pressure from reservoir which is equal to 50,000 kN per unit length. The factors of safety against sliding are evaluated using Eq. 3 for the four cases as shown in Fig. 4. It can clearly be seen that the evaluation of sliding safety factor based on the proposed PFOS leads to a conclusion that if we go for higher confidence levels on the shear strength parameters, there may be a possibility of failure of design ($F_S < 1$). However, using the PFOS based on IS 6512 [9] may be misleading specially if the uncertainty in the in-situ data is higher. Now, it is a question of engineering judgement whether

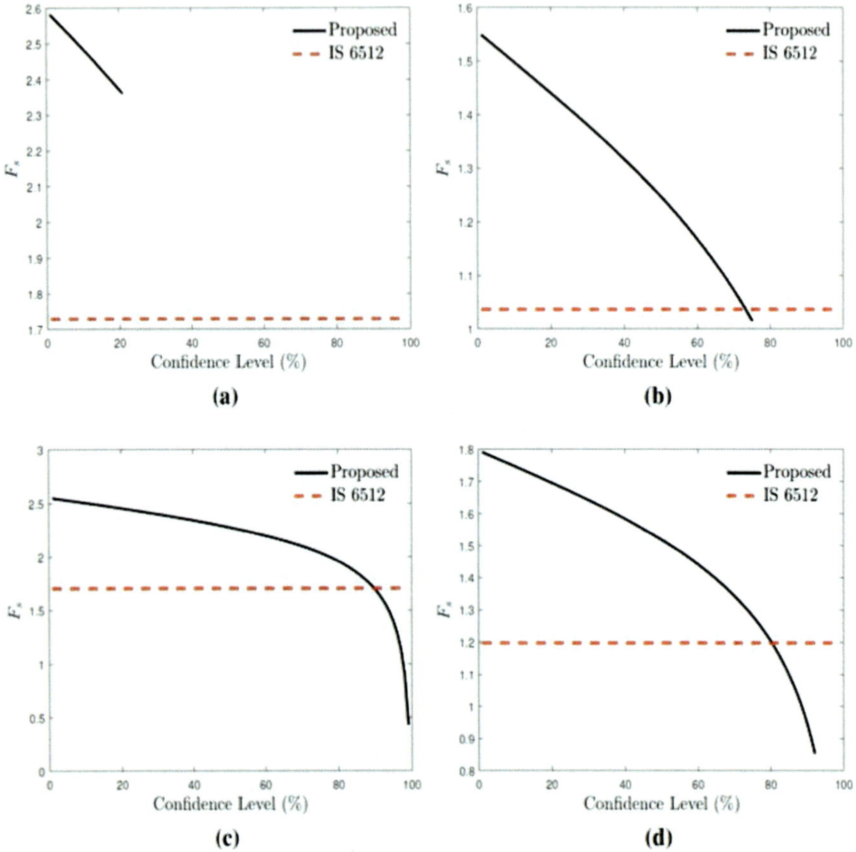

Fig. 4 Variation of factor of safety against sliding at **a** R/R and **b** C/R interfaces in weathered Phyllites rock-mass in Amouchu and **c** C/R interface in Garnetiferous-quartzo-feldspathic-biotite-gneiss rock-mass in Eastern Ghat Belts (India), and **d** C/R interface in Granite-gneiss rock-mass in Western Himalayas

stakeholders wish to make the design safe at all confidence levels or upto a certain level of confidence only. A better solution for the designers would be to employ the partial safety factors based on both the approaches (IS 6512 [9] and the proposed one) and then based on their choice of confidence level on in-situ data, they should investigate the safety of their designs.

4 Conclusions

The sliding stability of water or ground retaining structures depends upon the sliding resis- tance provided by cohesion and friction available at the structure-geomaterial or geomaterial- geomaterial interface. Therefore, a reliable estimation of in-situ shear strength parameters (cohesion and friction-angle) is of paramount importance. In case of rock-masses, these pa- rameters are evaluated by fitting Mohr–Coulomb criterion on in-situ direct shear test data and then to account for various physical and environmental uncertainties, partial factors of safety are applied on the two shear strength parameters as suggested by IS 6512. However, IS 6512 does not lay down any explicit procedure to account for the uncertainties associated with: (i) the assumption of linearity associated with Mohr–Coulomb criterion, (ii) geological features of rock-mass, and (iii) uncontrolled testing errors. In view of this, an explicit and simple procedure is proposed to assess the efficacy of IS 6512 to account for the testing data related uncertainties. The proposed procedure defines a site-specific partial factor of safety. Four different case studies are considered to illustrate the implications of the proposed safety factor and it is observed that at high level of confidence on the strength parameters, the partial safety factor of IS 6512 may not remain adequate and may lead to an unsafe design if testing uncertainties are higher. The results shown in this study encourage the practising professionals to use the proposed procedure along with conventional IS 6512 based procedure to asses the safety of their designs.

References

1. Alejano LR, Bobet A (2012) Drucker–prager criterion. In: The isrm suggested methods for rock characterization, testing and monitoring: 2007–2014, pp 247–252, Springer
2. Alonso EE, Pinyol NM, Pineda J (2014) Foundation of a gravity dam on layered soft rock: shear strength of bedding planes in laboratory and large “in situ” tests. Geotech Geol Eng 32(6):1439–1450
3. Barton N (2013) Shear strength criteria for rock, rock joints, rockfill and rock masses: problems and some solutions. J Rock Mech Geotech Eng 5(4):249–261
4. Brown E, Hoek E (1980) Underground excavations in rock. CRC Press
5. Chen Z, Xu J, Sun P, Wu C, Wang Y, Chen L (2012) Reliability analysis on sliding stability of gravity dams: part i, an approach using criterion of safety margin ratio. J Hydroelectr Eng (3):27
6. Eberhardt E (2012) The hoek–brown failure criterion. Rock Mech Rock Eng 45(6):981–988
7. Hao N, Li X (2020) Reliability analysis for the surface sliding failure of gravity dam. In: The international conference on embankment dams, pp 283–288
8. International Commission on Large Dams (ICOLD) (1995) Dam failures statistical analysis
9. IS 6512 (1984) Indian standard criteria for design of solid gravity dams (reaffirmed 1998). Bureau of Indian Standard, New Delhi
10. Kreuzer H, Bury K (1991) Safety assessment of concrete dams: safety factor versus reliability index. Dam Eng 2(2):101–116
11. Krounis A, Johansson F, Spross J, Larsson S (2017) Influence of cohesive strength in probabilistic sliding stability reassessment of concrete dams. J Geotech Geoenviron Eng 143(2):04016094

12. Labuz JF, Zang A (2012) Mohr–coulomb failure criterion. In: The isrm suggested methods for rock characterization, testing and monitoring: 2007–2014, pp 227–231, Springer
13. Lin T, Zhang J (1981) Development of the strength theory for rocks at the last decade in chinese. Mech Pract 3:17–23
14. Montgomery DC, Peck EA, Vining GG (2001) Introduction to linear regression analysis. Wiley
15. Pathak S, Ramana G, Dev H, Singh R (2015) Probabilistic interpretation of in-situ shear test data. In: Proceedings of 50th indian geotechnical conference (college of engineering, pune, india, December 2015)
16. Ramana GV, Pathak S, Dev H (2019) Shear strength parameters of granite rock mass: A case study. In: Stalin V, Muttharam M (eds), geotechnical characterisation and geoenvironmental engineering: lecture notes in civil engineering (Indian geotechnical conference-2016, IIT Madras), Vol 16, pp 273–280, Springer, Singapore
17. Ramana GV, Pathak S, Kumar N (2013) Selection and interpretation of shear strength parameters for weak phyllites. In: Kwasniewski, Łydzba (eds), rock mechanics for resources, energy and environment: ISRM–international symposium: EUROCK-2013, pp 331–336, Taylor & Francis Group, Wroclaw, Poland, London
18. Sheorey PR (1997) Empirical rock failure criteria. AA Balkema
19. Singh R (2009) Measurement of in situ shear strength of rock mass. J Rock Mech Tunn Technol ISRMTT 15(2):131–142
20. Yu M-H, Zan Y-W, Zhao J, Yoshimine M (2002) A unified strength criterion for rock material. Int J Rock Mech Min Sci 39(8):975–989
21. Zhou Z, Zhou F, Pan J-x, Wang Z-y (2008) Study on stability against deep sliding of gravity dam. Rock and Soil Mech 6

Mechanical Behaviour of Bentonite-Cement Mixtures Subjected to Change in Moisture Content

Rakesh Kumar Dutta, Jitendra Singh Yadav, and Ambuj Kumar Shukla

Abstract The expansive soils are subjected to change in volume due to variation in water content which results in problems such as high compressibility, reduced bearing capacity, differential settlement etc. Motivated by the issue, this experimental study aims to evaluate the mechanical behavior of bentonite- cement mixtures subjected to change in moisture content. Compaction, Atterberg limits and unconfined compressive strength tests were conducted to assess the influence of cement content (2, 4, 6, 8 and 10% by dry weight of soil), water content ($\omega_{opt}-3\%$, ω_{opt} and $\omega_{opt}+3\%$) and curing period on the mechanical behaviour of the bentonite soil. The results revealed that with addition of cement, the maximum dry unit weight of the bentonite increased and optimum water content decreased. A continuous reduction the liquid limit was observed with the inclusion of cement. Whereas, up to 6% incorporation of cement had led to increment in plastic limit of bentonite. The upsurge in the strength of bentonite was observed with addition of cement and prolongation curing period. The rise in strength of bentonite-cement specimens prepared at $\omega_{opt}+3\%$ were higher than that of specimens prepared at ω_{opt} and $\omega_{opt}-3\%$ water content.

Keywords Bentonite · Cement · Compaction · Atterberg limits · Strength

1 Introduction

In today's world where infrastructure is the key instrument to carry out the development forward, one needs to bring out the innovation and technology to benefit and speed up such progress. One such innovation, soil stabilization technique which has been prevalent since World War II, needs to be further refined and brought in a more precise manner so as to suit the suitability of the work. Soil stabilization becomes absolutely necessary where the ground condition is not fit enough to continue with the construction work over and hence its engineering properties are modified using

R. K. Dutta · J. S. Yadav (✉) · A. K. Shukla
Department of Civil Engineering, National Institute of Technology Hamirpur, Hamirpur, H.P, India
e-mail: jitendershine@gmail.com

A. Garg et al. (eds.), *Proceedings of the 1st Indo-China Research Series in Geotechnical and Geoenvironmental Engineering*, Lecture Notes in Civil Engineering 123,
https://doi.org/10.1007/978-981-33-4324-5_10

certain method or technique. Out of total India's soil cover area, approximately 20% constitutes 'expansive soil' which has the peculiar characteristic of showing a large increase in its volume on coming in contact with water. Such soils are said to be unstable and unfit for construction purpose and hence comes the role of soil stabilization.

Various methods of stabilization have been devised till date and the most used methods are based on the principles such as mechanical, chemical, use of cement, lime or any other additive, bituminous employment, etc.

Cement is the oldest and most preferred stabilizing agent due to its strong binding properties. Cement reaction does not depend upon the soil minerals and it can be used with any type of soil. A wide variety of cement is available in the market and its choice depends upon the type of soil and desired degree of strength required. It is done by blending pounded soil and Portland cement with water and compacting the mix to attain a strong soil base. As the cement hydrates it imparts strength to the mix. A large amount of cement is produced in the country on daily basis and due to its easy availability, it is still the most popular stabilizer even after its few short-comings.

Cement and lime has been most commonly used stabilizers by many investigators from last few decades [1, 2, 4, 5, 8, 13, 15, 17, 18, 21, 23]. The reduction in liquid limit, plasticity index and swelling potential, and increment shrinkage limit and shear strength of cohesive soil were noticed by the addition of cement [3, 11, 12, 20, 22]. The behavior of cohesive soil with incorporation of cement depends upon cation exchange, flocculation and agglomeration, cementitious hydration, and pozzolanic reaction processes [16]. Kim et al. [10] evaluated effect of bentonite and cement on permeability and strength behavior of soil liner. They recommended 5% bentonite and 5% cement as an optimal dose to be mix with clayey soil. Lime and Fly Ash were added to an expansive soil by Zhang and Cao [9] in the percentage range of 4% to 6% and 4–0% to 50% by dry weight of the soil and it was seen that as the proportion of lime and fly ash went on to increase there was a mark reduction in the value of maximum dry density and a corresponding increase in OMC (%) along with increase in the proportion of coarse particles. Plastic limit increased due to mixing of lime and liquid limit decreased by mixing fly ash, which ultimately decreased the plasticity index. Xiao and Lee [21] assessed the effects of curing period on the stress-strain characteristic of cement treated marine clay. They reported that curing time had a substantial impact on the unconfined compressive strength, isotropic compression behaviour and stress–strain behaviour. Okeke et al. [14] examined the effect of increase in cement quantity on the strength and compaction parameters of some lateritic soils. Results showed that stabilization of the soils with cement increased its UCS. Furthermore, addition of cement beyond 10% by weight caused reduction in UCS.

There is limited number of literature available on the mechanical behavior of bentonite-cement mixtures subjected to variation in moisture content. Thus, this study reports the mechanical behavior of bentonite amended with cement, intending to evaluate the compaction, Atterberg limits and strength at different water contents and curing durations, directing on engineering projects applications.

Table 1 Physical characteristics of bentonite

Attributes	Value
Specific gravity	2.3
LL (%)	374.79
PL (%)	52.41
Optimum Moisture content, ω_{opt} (%)	38.43
Maximum dry unit weight, γ_{dmax} (kN/m^3)	12.39
Soil form	CH

Table 2 Mineralogical composition of bentonite

Minerals	Percentage composition (%)
Oxygen (O)	39.34
Carbon (C)	26.21
Silica (Si)	12.16
Iron (Fe)	11.24
Aluminium (Al)	7.12
Barium (Ba)	2.92
Sodium (Na)	0.76
Magnesium (Mg)	0.25

2 Materials

In this investigation Sodium bentonite procured from local market was used. The physical characteristics of the bentonite is tabulated in Table 1. The Atterberg limits of the bentonite are liquid limit 374.79% and plastic limit 52.41%. The specific gravity of solids is 2.3. According to the IS 1498 (1970) [6], the soil can be classified as a clay of high plasticity (CH). The mmineralogical composition of bentonite is shown in Table 2. Scanning electron microscopy image shows that the particles of bentonite are flaky in nature (Fig. 1b). Ordinary Portland Cement of Grade 43, manufactured by ACC Cement Limited, was used in this investigation.

3 Experimental Program

In this investigation, the bentonite was mixed with 2%, 4%, 6%, 8%, and 10% cement by the dry weight bentonite. Mixtures were named according to their cement content: B0 for pure bentonite; B2 for 2% cement-98% bentonite; B4 for 4% cement-96% bentonite; B6 for 6% cement-94% bentonite; B8 for 8% cement-92% bentonite and B10 for 10% cement-90% bentonite. The mixtures of bentonite and varying cement content were subjected to series of compaction, consistency limits and shear strength tests followed by microstructural analysis.

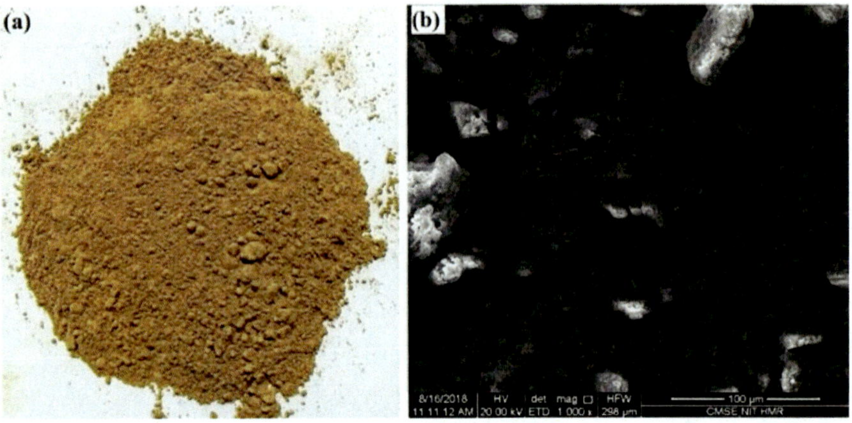

Fig. 1 **a** Bentonite used; **b** SEM of bentonite (1000*x*)

The compaction tests were carried out in a mini compaction mould (100 mm in length and 38 mm in diameter) as suggested by Sridharan and Sivapullaiah [19]. As it requires only one-tenth (1/10) of the volume of soil required in Proctor test and saves effort and time as well. The compaction parameters (γ_{max} and ω_{opt}) for each mixture's cement content were determined. The Consistency (Atterberg) limits i.e. liquid limit (W_L) and plastic limit (P_L) of bentonite-cement mixtures were determined in accordance with IS 2720 (Part V) 1970. Before subjected to the tests, the mixtures were mixed with desired ω_{opt} and stored in an air tight packet for 0, 1, 7, and 28 days of curing. The unconfined compressive strength (UCS) of bentonite-cement specimens of 38 mm diameter and 76 mm height prepared at $\omega_{opt} - 3\%$, ω_{opt}, and $\omega_{opt} + 3\%$ were determined in agreement with IS 2720-Part-X (1991) [7]. The specimens were cured for 0, 1, 7, and 28 days at 96% humidity and 27 °C temperature. A total of 216 specimens were prepared and tested. In order to get the better intrusion of the micro-structural behavior of bentonite-cement mixtures cured for different periods, Scanning Electron Microscopy was carried out. A small portion of the tested UCS samples cured for different periods were dried and kept in in air tight bags.

4 Results and Discussions

4.1 Compaction Characteristics

The variation of γ_{max} and ω_{opt} for different bentonite—cement mixtures are shown in Fig. 2. An increase in γ_{max} of the bentonite with an increase in cement content was observed. The γ_{max} values were 12.39 kN/m^3, 12.47 kN/m^3, 12.62 kN/m^3, 12.70 kN/m^3, 12.88 kN/m^3, and 12.95 kN/m^3 for B0, B2, B4, B6, B8, and B10 samples, respectively. These observations are in agreement with Tatsuoka et al. [20]. The

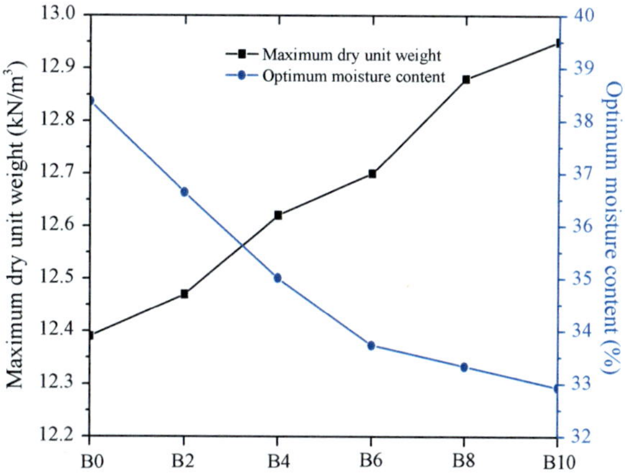

Fig. 2 Variation of γ_{max} and ω_{opt} of bentonite at varying percentage of cement

increase in γ_{max} can be apprehended to the binding property of cement which attracts all the soil particles to its surface and binds them together. Thus countless cement particles attract soil particles and hold them together resulting in decrease in void ratio which ultimately causes increase in density index. The mechanisms responsible for this process are higher specific gravity of cement, Base Cation Exchange and Particle Restructuring. Base Cation Exchange process causes the replacement of sodium ion from bentonite with the calcium ion supplied by cement and Particle Restructuring causes the particle to change from more plastic and fine grained to friable and coarse grained which is an alteration to the texture and thus the dry density also increases. Cement hydration causes the formation of two compounds calcium-silicate-hydrate and calcium-aluminium-hydrate which act as glue and help in binding of particles causing increase in γ_{max}.

In contrast, the decrease in ω_{opt} of bentonite is observed with the increase in cement content. The ω_{opt} values were 38.43%, 36.68%, 35.07%, 33.81%, 33.33%, and 32.93% for B0, B2, B4, B6, B8, and B10 samples, respectively as shown in Fig. 3b. The reduction of ω_{opt} may be associated with the release of water trapped in the flocs as the particles started to attain coarse shape which may reduce the surface area during compaction. The excess of water gets utilised in the hydration of cement and the compaction process becomes a little easy.

4.2 *Consistency (Atterberg) Limits*

The impact of cement inclusion on the liquid limit (W_L) and plastic limit (P_L) of bentonite are shown in Fig. 3a–b. A clear decrease in W_L of bentonite is observed with the increase in cement content as illustrated in Fig. 3a. At 0 day of curing, the W_L values were 374.79%, 330.95%, 224.65%, 202.73%, 201.64% and 195.15% for B0, B2, B4, B6, B8, and B10 samples, respectively. The reduction in W_L for all

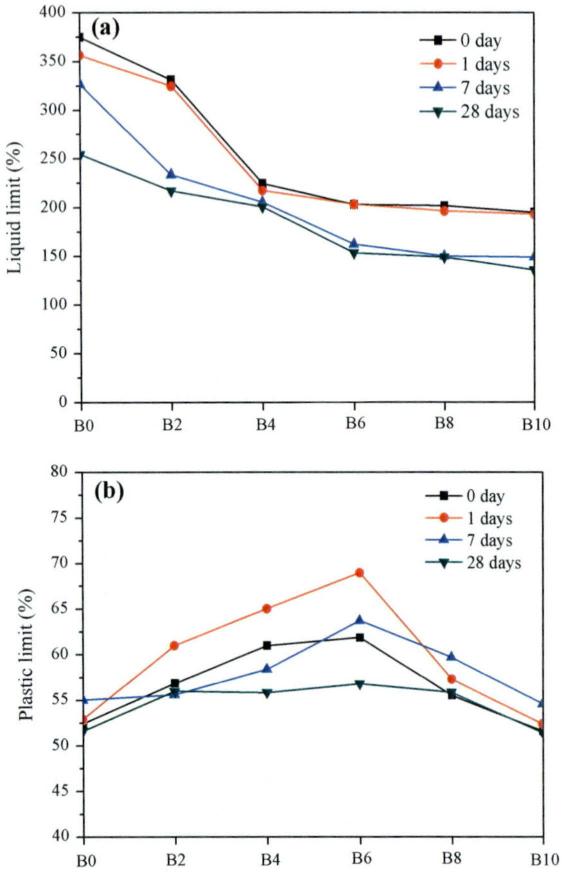

Fig. 3 Variation in (**a**) liquid limit, (**b**) plastic limit of bentonite-cement mixtures with age

mixtures is associated to the change in bentonite structure due to addition of cement. The binding nature of cement attracts the bentonite particles together and makes them coarse enough to reduce the liquid limit. With the prolongation of curing period, the reduction in W_L of bentonite-cement mixtures are observed. At 28 days of curing, it decreased to 254.24%, 216.98%, 200.54%, 153.42%, 149.10% and 135.98% for B0, B2, B4, B6, B8, and B10 samples, respectively. This reduction in W_L with prolongation of curing period may be attributed to the evolution of heat of hydration with time, which ultimately imparts strength and makes the mixture less plastic.

The plastic limit obtained for bentonite-cement mixtures are presented in Fig. 3b. It is seen that P_L of the bentonite increases with inclusion of cement up to 6%, after that it fall down. The plastic limit values for 7 days cured samples of B0, B2, B4, B6, B8, and B10 were 54.96%, 55.53%, 58.34%, 63.69%, 59.68% and 54.55%, respectively. The increase in P_L can be attributed to deficiency of water for the heat of hydration. However, the fall in P_L after 6% cement content might be due to development of coarse grained particles as a result of binding action of the cement. Figure 3b reveals that the effect of curing on the P_L of pure bentonite and B10 samples is insignificant. The value remained almost the same for all the mixtures. However, the change in P_L of bentonite-cement mixtures cured for 7 and 28 days was not very noteworthy and it remained very close to each other. Overall, the addition of cement does negate the plastic characteristics of bentonite to some extent and helps in reducing the plasticity index.

4.3 Unconfined Compressive Strength

Figure 4a–c show the axial stress–strain curves of the specimens of bentonite–cement mixtures prepared at $\omega_{opt} - 3\%$, ω_{opt}, and $\omega_{opt} + 3\%$ and cured for 28 days. It can be inferred from Fig. 4a–c that the axial stress of bentonite specimen increases significantly with the increase in cement content and curing period. The formation of chemical bond between cement and bentonite particles, which lead to better interlocking and densification may be responsible for the increase axial stress. Moreover, the specimens prepared at $\omega_{opt} + 3\%$ exhibited greater axial stress and failure strain compared to that of specimens prepared at $\omega_{opt} - 3\%$ and ω_{opt} at the specific cement content and curing period. The failure pattern of bentonite-cement mixture specimen is illustrated in Fig. 5. The specimen fails along the shear plane at angle approximately 45° with longitudinal loading plane, demonstrating brittle failure.

The impact of variation in ω_{opt} on the 0, 1, 7, and 28 days cured specimens of bentonite-cement mixtures are shown in Fig. 6a–c. At ω_{opt} -3% water content, an increase in strength of bentonite-cement specimens is observed with the increase in cement content up to 8%. The UCS values were 1245.3 kN/m^2, 1375.56 kN/m^2, 1515.56 kN/m^2, 1575.54 kN/m^2 and 1635.56 kN/m^2 for B0, B2, B4, B6, and B8 samples, respectively. Thereafter, at 10% inclusion of cement a down fall is witnessed i.e. 1405.98 kN/m^2. The reduction in strength for B10 mixtures can be attributed to less availability of water for hydration reaction. With prolongation of curing period from 0 to 28 days, an increase in the UCS of specimens is perceived with can be accredited to formation of cementitious compounds to achieve formidable strength.

At ω_{opt}, as the cement content increases the UCS of bentonite-cement mixtures increases (Fig. 6b). For example, the UCS values for B0 were 191.51 kN/m^2, 282.63 kN/m^2, 802.45 kN/m^2 and 1313.12 kN/m^2 at curing period of 0, 1, 7, and 28 days which increased to 291.84 kN/m^2, 656.23 kN/m^2, 1340.52 kN/m^2 and 2343.54 kN/m^2 for B10 mixtures at the same age. The formation of C–S–H and C–A–H compound in the mixtures which provides the binding action to the cement may be possible reason for the observed behaviour. With the prolongation in curing period, an increase in strength of bentonite-cement mixtures is seen. At the 7th days UCS values of B0, B2, B4, B6, B8, and B10 specimens were 802.45 kN/m^2, 785.96 kN/m^2, 1015.82 kN/m^2, 844.45 kN/m^2, 928.34 kN/m^2 and 1347.52 kN/m^2 which increased to 1313.24 kN/m^2, 1752.60 kN/m^2, 1872.95 kN/m^2, 1984.25 kN/m^2, 2305.2 kN/m^2 and 2343.54 kN/m^2 for the same specimens after an age of 28 days. The increase in strength with age is associated to release of heat of hydration with is in the range of 80–90 cal/g at 7 days

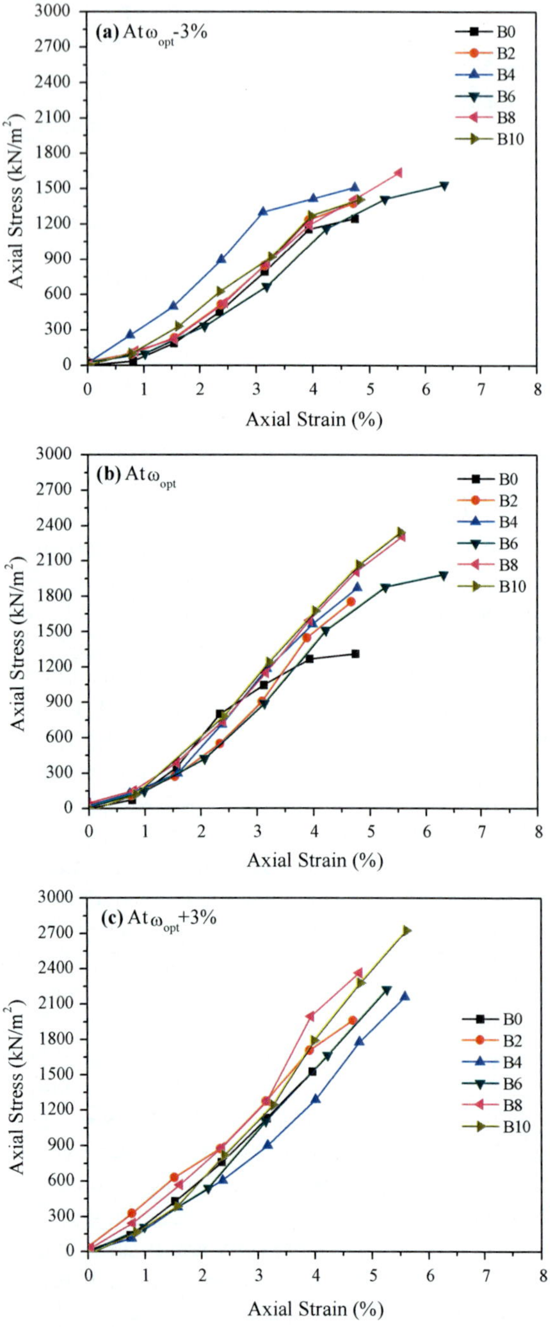

Fig. 4 Axial stress and strain curves of bentonite-cement mixtures cured for 28 days **a** $\omega_{opt}-3\%$, **b** ω_{opt}, **c** $\omega_{opt}+3\%$

Fig. 5 Failure specimen of bentonite-cement mixtures

and increases to 90–100 cal/g after 28 days. Similar to the other water content, at $\omega_{opt} + 3\%$ water content, an upsurge in the UCS of bentonite-cement mixtures is noticed with rise in dose of cement and age. The study of Fig. 6a–c reveal that strength of bentonite-cement mixtures prepared at $\omega_{opt} + 3\%$ water content is higher compared to that of $\omega_{opt} - 3\%$ and ω_{opt} at the same cement content and curing period. Similar the strength specimens prepared at ω_{opt} is greater than that of specimens prepared at $\omega_{opt} - 3\%$. For example, at $\omega_{opt} + 3\%$ water content, the UCS values at 28 days were 1597.18 kN/m^2, 1959.51 kN/m^2, 2158.49 kN/m^2, 2223.69 kN/m^2, 2361.02 kN/m^2 and 2723.65 kN/m^2, whereas at ω_{opt} and $\omega_{opt} - 3\%$ water content, UCS values were 1313.24 kN/m^2, 1752.60 kN/m^2, 1872.95 kN/m^2, 1984.25 kN/m^2, 2305.2 kN/m^2 and 2343.54 kN/m^2; and 1245.3 kN/m^2, 1375.56 kN/m^2, 1515.56 kN/m^2, 1575.54 kN/m^2, 1635.56 kN/m^2 and 1405.98 kN/m^2, respectively for B0, B2, B4, B6, and B8 specimens at same age. The availability of sufficient water to satisfy the heat of hydration of cement which benefits in attaining the strength may be reason for maximum increase in strength at $\omega_{opt} + 3\%$ water content.

4.4 SEM Analysis

Figure 7a–c show the microstructural images of the bentonite-cement specimens. The presence of void can easily be seen in Fig. 7a for B0 specimen. As can be seen in Fig. 7b, the inclusion of cement in the bentonite has led to formation of C–S–H compound which resulted in agglomeration of the particles and change in gradation of bentonite. It resulted into compaction and densification, which is the for the increase in the strength. Figure 7b exhibits the existence of less quantity of C–S–H compounds in the B8 specimens which is due to less availability of water for cement to satisfy its heat of hydration.

Fig. 6 Variation in UCS of bentonite-cement specimens with curing period **a** $\omega_{opt} - 3\%$; **b** ω_{opt}; **c** $\omega_{opt} + 3\%$

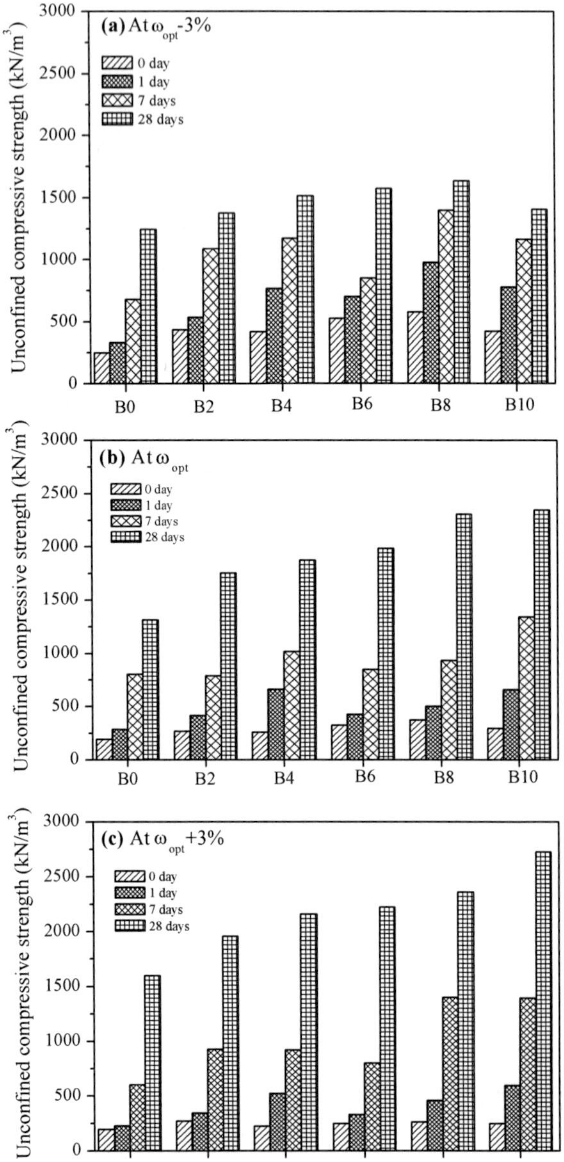

Fig. 7 Microstructural images of specimens cured for 28 days **a** B0 at 200 × (ω_{opt}), **b** B4 at 200 × (ω_{opt} + 3%), **c** B8 at 1000× ((ω_{opt} − 3%)

5 Conclusions

Based on the results and analyses presented on the paper, the conclusions derived are as follows:

- Compaction tests conducted on various bentonite-cement mixtures showed an increasing γ_{max} and decreasing ω_{opt} with increasing cement content.
- A declining trend in W_L of various bentonite-cement mixtures with increasing cement content and curing period was observed. Whereas, P_L of bentonite-cement mixtures showed increasing trend up to 6% inclusion of cement. Beyond that it started to form slump and resulted in decreasing P_L.
- An increasing UCS of bentonite-cement mixtures with increasing cement content and curing period was observed.
- UCS of bentonite-cement mixtures specimens prepared at $\omega_{opt} + 3\%$ were higher as compared to the same mixtures prepared at ω_{opt} and $\omega_{opt} - 3\%$ water content.

References

1. Azadegan O, Yaghoubi MJ, Pourebrahim GR (2010) Effect of completely dried materials in natural condition on mechanical properties of lime/cement treated soils. Electron J Geotech Eng 15:1727–1736
2. Azadegan O, Yaghoubi MJ, Pourebrahim GR (2011) A laboratory study of the behavior of the lime/cement slurry and compacted un-reinforced piles. Electron J Geotech Eng 16:375–386
3. Bahar R, Benazzoug M, Kenai S (2004) Performance of compacted cement-stabilised soil. Cement Concrete Comp 26(7):811–820
4. Chiu CF, Zhu W, Zhang CL (2009) Yielding and shear behaviour of cement-treated dredged materials. Eng Geol 103(1–2):1–12
5. Hossain KMA, Lachemi M, Easa S (2007) Stabilized soils for construction applications incorporating natural resources of Papua New Guinea. Resour Conserv Recycl 51(4):711–731
6. Indian Standard, I. S. 1498 (1970) Classification and Identification of soils for general Engineering purposes. Reprint, Bureau of Indian Standards, India
7. Indian Standard, I. S. 2720 (Part X)-1991 (2010) Methods of test for soils: Part 10 Determination of unconfined compressive strength
8. Jauberthie R, Rendell F, Rangeard D, Molez L (2010) Stabilisation of estuarine silt with lime and/or cement. Appl Clay Sci 50(3):395–400
9. Ji-ru Z, Xing C (2002) Stabilization of expansive soil by lime and fly ash. J Wuhan Univ Technol Mater Sci 17(4):73–77
10. Kim S-M, Youm H-N, Lim N-W (2000) Effect of bentonite and cement on permeability and compressive strength of the compacted soil liner. J Korean Soc Environ Eng 22(3):495–504
11. Lee F-H, Lee Y, Chew S-H, Yong K-Y (2005) Strength and modulus of marine clay-cement mixes. J Geotech Geoenviron Eng 131(2):178–186
12. Nelson J, Miller DJ (1997) Expansive soils: problems and practice in foundation and pavement engineering. Wiley
13. Ninov J, Donchev I, Lenchev A, Grancharov I (2007) Chemical stabilization of sandy-silty illite clay. J Univ Chem Technol Metall 42(1):67–72
14. Okeke OC, Okogbue CO, Imasuen OI, Aghamelu OP (2015) Engineering Behaviour of Cement-Treated Expansive Subgrade Soils from Awgu, Southeastern Nigeria. Civil Environ Res 7(6)

15. Okyay US, Dias D (2010) Use of lime and cement treated soils as pile supported load transfer platform. Eng Geol 114(1–2):34–44
16. Prusinski JR, Bhattacharja S (1999) Effectiveness of Portland cement and lime in stabilizing clay soils. Transp Res Rec 1652(1):215–227
17. Quarcioni VA, Cincotto MA (2006) Optmization of calculation method for determination of composition of hardened mortars of Portland cement and hydrated lime made in laboratory. Construct Build Mater 20(10):1069–1078
18. Reid JM, Brookes AH (1999) Investigation of lime stabilised contaminated material. Eng Geol 53(2):217–231
19. Sridharan A, Sivapullaiah PV (2005) Mini compaction test apparatus for fine grained soils. Geotech Test J 28(3):240–246
20. Tatsuoka F, Uchida K, Imai K, Ouchi T, Kohata Y (1997) Properties of cement-treated soils in Trans-Tokyo Bay Highway project. Proc Institution Civil Eng Ground Improve 1(1):37–57
21. Xiao HW, Lee FH (2008) Curing time effect on behavior of cement treated marine clay. World Acad Sci Eng Technol 43(2008):71–78
22. Yin JH, Lai CK (19998) Strength and stiffness of Hong Kong marine deposits mixed with cement. Geotech Eng 29(1)
23. Yong RN, Ouhadi VR (2007) Experimental study on instability of bases on natural and lime/cement-stabilized clayey soils. Appl Clay Sci 35(3–4):238–249

Development of Landslide Early Warning Using Rainfall Thresholds and Field Monitoring: A Case Study from Kalimpong

Neelima Satyam and Minu Treesa Abraham

Abstract Landslides in hilly regions cause lives and property loss and are considered as highly destructive natural disasters. Urbanization of highlands due to population rise makes this geomorphological process highly risky. In India, the Himalayan belt is experiencing hazardous landslides every year and the major triggering factor for such landslides is rainfall. Hence a satisfactory method for risk reduction is the development of Landslide Early Warning System (LEWS), which can help in issuing alert to the public regarding any possible landslides. This study explores in detail the different landslide forecasting methods for Kalimpong town in Darjeeling Himalayas, a highly susceptible landslide zone. Multiple rainfall thresholds are defined for the study area and have been validated using the real time filed monitoring observations using Micro Electro Mechanical Systems (MEMS) tilt sensors installed in the region. It was observed that an algorithm-based approach, called SIGMA is the best suited approach among the different rainfall thresholds. SIGMA can be used to issue multiple levels of warning based on the severity of landslides. The rainfall threshold can be used as the first line of action and warnings can be issued after verifying the field monitoring data. Thus, the combination of rainfall threshold and filed monitoring can be used to develop an efficient LEWS for Kalimpong.

Keywords Rainfall · Thresholds · Landslides · Kalimpong · Monitoring

1 Introduction

Landslides result in terrible life and property loss in hilly terrains. The destructions caused by such catastrophic events are increasing every year. The population rise, urbanization and the recent climatic changes have increased the number of landslides

N. Satyam (✉) · M. T. Abraham
Discipline of Civil Engineering, IIT Indore, Indore 453552, Madhya Pradesh, India
e-mail: neelima.satyam@gmail.com

M. T. Abraham
e-mail: minunalloor@gmail.com

A. Garg et al. (eds.), *Proceedings of the 1st Indo-China Research Series in Geotechnical and Geoenvironmental Engineering*, Lecture Notes in Civil Engineering 123,
https://doi.org/10.1007/978-981-33-4324-5_11

and the casualties associated with them. In India, rainfall is the primary triggering factor of initiation of landslides and most severe cases occur during the monsoon season. Landslides in India are often associated with extreme rainfall events. Hence most of the landslides occur along with other disasters and the impact is often underestimated. But in most cases of combined hazards, the losses from landslides may exceed all other losses [1]. The investments in risk reduction approaches can create significant impact in minimizing the loss due to landslides. With this insight, development of risk reduction strategies including Landslide Early Warning Systems (LEWS) are being attempted across the world [2–4]. Operational LEWS are very less in India and the development of such a system for high risk areas has been a rising concern for the authorities after the recent increase in land-slide disasters across the country.

An early warning system is expected to provide an intervention delay for making decisions and take necessary actions. This time can be used to reduce the risk. LEWS is a system which includes the awareness given to public, action plans, risk identification etc. The failure of any of these elements can make the whole system fail. For developing an LEWS, a strategy is fixed first and then conceptually modified with time. In India, Himalayas and Western Ghats [5–7] are the most landslide susceptible zones. This study discusses in detail the development of a landslide forecasting system and its evaluation using real time field monitoring, for Kalimpong town in India.

Kalimpong is in the state of West Bengal in India and is highly affected by rainfall induced landslides. The fatalities due to landslides have both direct and indirect effects on the population. This famous tourist destination of West Bengal is facing severe loss in transportation and infrastructure sectors due to landslides every year. A detailed analysis has been conducted to find out the best suited landslide forecasting model for the study area and the performance has been evaluated using real time field monitoring data. Subsequent paragraphs, however, are indented.

2 Study Area

Kalimpong in Darjeeling Himalayas is a well-known tourist destination. This hilly terrain with elevation upto 1650 m (Fig. 1) lies between two rivers viz. Relli (east) and Teesta (west). The catchment basin of Teesta river contains steep rugged slopes while the eastern slopes are gentle. The climatic and topographical features of this region make it a common destination and hence tourism and agriculture are the major income sources of the locale. During monsoon season, slope failures are often triggered by incessant rainfalls in the western slopes of Kalimpong town. Such incidents disrupt the transportation facilities and isolates the town.

Phyllite, schists and archean gneiss contributes to the geological formation of Kalimpong [9]. The bedrock, across the region, consists of quartz mica schist of golden to silver color which belongs to Daling series [10]. The rock, in general, is metamorphosed up to the biotite zone, and near the eastern margin, garnet starts

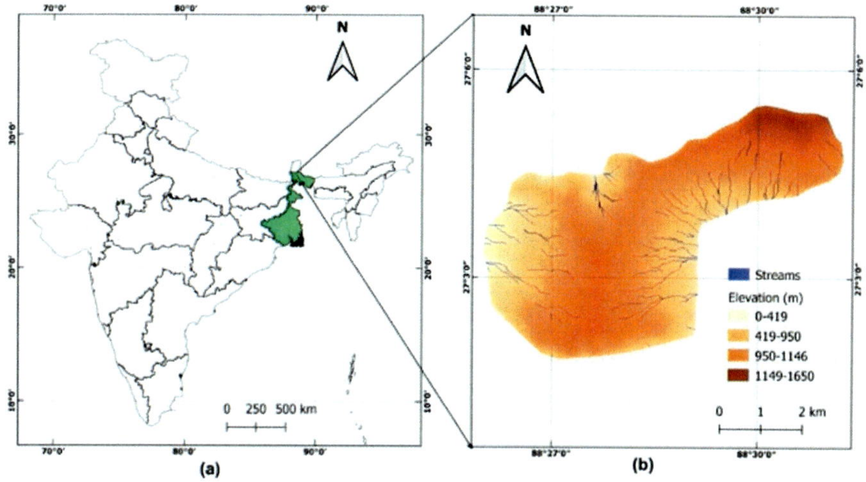

Fig. 1 Location, elevation [8] and drainage details of Kalimpong

appearing. The inclination varies from nearly 20° at the area near to Teesta to 40° towards town. The joints present in rocks are generally found in three directions: one parallel, one perpendicular and also oblique to the foliation. Foliation of schist and enrichment of micaceous minerals control the occurrence of landslides in the town to a great extent. Jointed rocks and cracks lead to disintegration and decomposing of rocks to form unstable matter. The organic soils with very high moisture holding capacity leads to expansion of soil volume during rains. The rain water percolating continuously till the bottom layer leaves the middle layer with relatively coarser particles and in turn reduces the shear strength of soil [11].

The major streams of the region flow towards Teesta river [10]. The lower order streams are often discontinuous, they disappear at one point and later evolve at downstream. Morphometrically, the slope ranges from gentle slope category to steep slope category [10]. During monsoon season, the streams along slopes flow in very high velocities, resulting in floods and slope failures in the region. The erosion along banks of streams advances to bank failures and critical slope failures. The rainfall data of Kalimpong, during monsoon seasons from 2010 to 2016 is tabulated below, in Table 1.

The highest annual rainfall obtained during the study period was 2061 mm. The rainwater passes through the cracks and rock joints, leading to landslides. The continuous percolation makes the rocks weak and accelerates weathering of rocks. Hence to develop an effective LEWS, first step is to understand the effect of the major triggering factor, landslides.

Table 1 Daily precipitation (mm) during monsoon season for Kalimpong town (2010–2016) [12]

Month/Year	June	July	August	September
2010	316.8	665.4	425.3	268.2
2011	337.0	678.0	525.6	384.1
2012	354.9	433.1	250.8	467.9
2013	248.0	424.6	401.0	113.0
2014	396.4	371.2	571.8	265.4
2015	568.0	534.4	242.3	331.2
2016	327.2	869.8	262.6	366.8

3 Rainfall Thresholds

A critical condition of the system, beyond which a state of change is expected is called a threshold. In case of landslides, it is often critical levels of meteorological or hydrological parameters beyond which landslides are expected to occur [6, 13–16]. In real cases, thresholds do not separate the state of system into these ideal parts, but there can be a minimum threshold value which is the level below which landslides will never occur and a maximum threshold level beyond which landslides always happen [16]. Minimum threshold levels are associated with too many false alarms while maximum threshold levels can lead to many missed alarms. Sometimes, intermediate thresholds are used to define the states which lie between these minimum and maximum conditions.

3.1 Empirical Thresholds

Empirically based approaches uses the historical relationship between rainfall and landslides in the study area. By using the rainfall events that have triggered landslides in history, the lower bounds of rainfall parameters are identified and plotted in a cartesian, semi-logarithmic or logarithmic co-ordinates [16]. The rainfall parameters include the total cumulative rainfall from the beginning of continuous rainfall till the occurrence of landslide, known as Event rainfall (E), the duration of rainfall which is the time duration of an event rainfall (D) and the intensity of rainfall (I) which is the rate of precipitation, calculated as the ratio of event to duration. The threshold equations are in the form of a power law and hence logarithmic plot is preferred for representing the multiple order data in the form of a straight line. The thresholds can be defined in ID, ED or EI planes. The pioneering of Caine in 1980 [13] on global ID thresholds was a milestone in the history of definition of ID thresholds across the world. The functional dependency of I and D is found to be the major limitation of this approach and the recent literature shows a shift towards ED thresholds. ED thresholds were first proposed in 1985 [17] and is being widely followed by practitioners around the globe [18–22]. All these thresholds are based on the parameters associated with the rainfall event responsible for the occurrence of landslides. Another category of empirical thresholds consider the effect of antecedent rainfall, instead of a specific rainfall event for the definition of threshold [14, 16].

The lower bounds of rainfall parameters were identified visually, without any statistical analysis in the earlier stages. The rainfall events that have triggered or not triggered landslides were separated using a minimum threshold line, by visual inspection [23, 24]. It was in 2007 that the lower bounds were first determined using a Bayesian inference approach [16] which was updated in 2010 using a Frequentist approach [15]. The Frequentist approach has got a wider acceptance and has been used for defining both ID and ED thresholds [5, 6, 25, 26]. The frequentist approach has been modified in 2012 by using a statistical bootstrapping, which has helped in estimating the uncertainty associated with the defined thresholds [18].

For Kalimpong, both ID and ED thresholds are defined using frequentist approach [14, 20, 22, 27]. The historical landslide and rainfall data from 2010 to 2013 have been used for the definition of thresholds and have been validated using the geotechnical monitoring data of 2017.

Intensity-Duration Thresholds for Kalimpong

The ID threshold for Kalimpong was defined using a frequentist approach [14] using the daily precipitation data collected from Geological Survey of India and the landslide records collected from multiple government agencies and media reports.

The threshold can be expressed in power law form as:

$$I = \alpha D^{-\beta} \tag{1}$$

where I is the average rainfall intensity in mm/h, D is the duration of rainfall event in hours, α is the scaling parameter and β is the shape parameter. The equation can be expressed in the form of a straight line as:

$$\log I = \log \alpha - \beta \log D \tag{2}$$

With the data plotted in logarithmic scale, the best fit line has been obtained using the least square approach. The distribution of data around the best fit line was then evaluated and the shift of each point in the y direction from the best fit line was fitted using a standard Gaussian function using a kernel density estimation. The equation of threshold with 5% exceedance probability was then derived as $I = 3.52\ D^{-0.41}$ [14] (Fig. 2). The probability that a landslide may occur is very less below this threshold. The threshold was found to be lower than the other rainfall thresholds defined across the world [14].

Antecedent Rainfall Thresholds for Kalimpong

Apart from the effect of single event rainfall, the possible effect of antecedent precipitation should also be considered as a triggering factor for landslides. From short

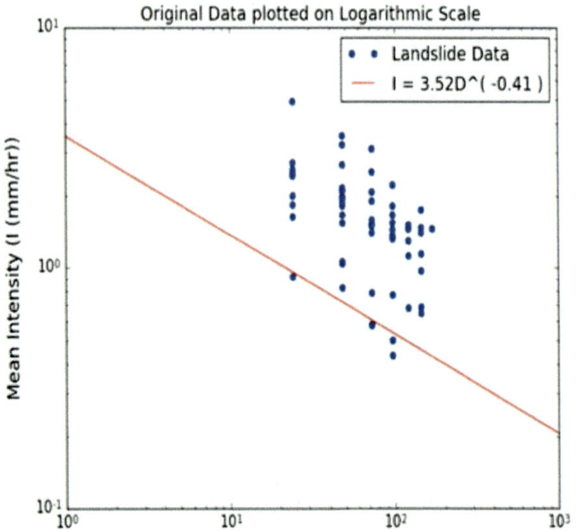

Fig. 2 Intensity-duration thresholds for Kalimpong [14]

durations (3 days) to long durations (30 days) were considered for the analysis. The biasness of initiation of landslides to daily/antecedent rainfall has been evaluated [14].

The analysis depicted that the occurrence of most landslides had a shift towards the antecedent rainfall than the daily rainfall (Fig. 3). Thresholds were defined for 10 day antecedent precipitation as 88.37 mm and for 20 days as 133.5 mm [14].

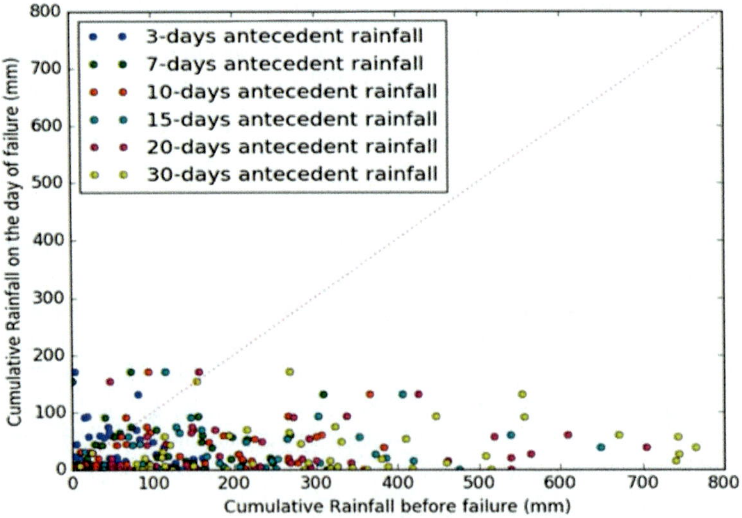

Fig. 3 Development of antecedent rainfall thresholds for Kalimpong [14]

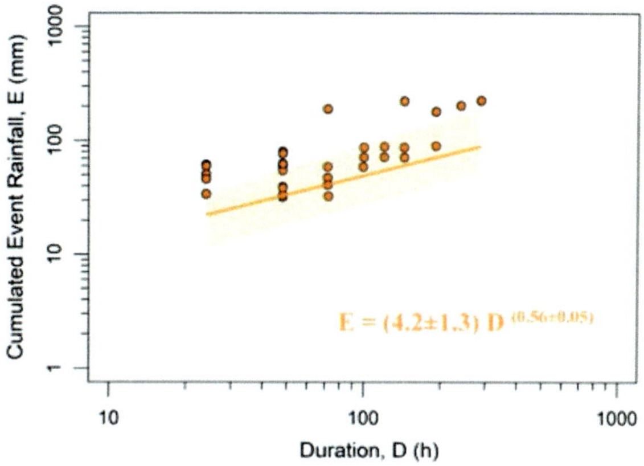

Fig. 4 Event-duration thresholds for Kalimpong [20]

Event-Duration Thresholds for Kalimpong

For defining ED thresholds, an algorithm proposed by Melillo et al. [19] has been used. The tool reconstructs the rainfall events in the study area and identifies the rainfall events responsible for occurrence of landslides after finding the maximum probability rainfall conditions (MPRC). The algorithm is called Calculation of Thresholds for Rainfall Induced Landslides—Tool (CTRL-T). It takes the location details of rain gauges and landslides, and the time series of rainfall and landslides as input. Based on the input details, the tool defines ED thresholds with multiple exceedance probabilities. The rainfall parameters associated with each MPRC is used to derive the threshold power law in the following format:

$$E = (\alpha \pm \Delta\alpha) D^{(\gamma \pm \Delta\gamma)} \quad (3)$$

where γ is the shape parameter α is the scaling parameter. The value of γ is given by $\gamma = 1 - \beta$ where β is the shape parameter of ID thresholds. The uncertainties in defining α and γ are denoted by $\Delta\alpha$ and $\Delta\gamma$. A non-parametric bootstrapping technique is used by the tool to calculate the uncertainty associated with each level of exceedance probability.

By using frequentist approach, thresholds at multiple levels are defined using a frequentist approach with Gaussian distribution as target function. The threshold with 5% exceedance probability has been defined as the minimum threshold as shown in Fig. 4.

3.2 *SIGMA Model*

SIGMA (*Sistema Integrato Gestione Monitoraggio Allerta*) model was developed for landslide earl warning in Italy [28]. As the name says, the model makes use of standard deviation of a statistical distribution to define thresholds. The model based on statistical analysis is easy to be exported to other regions as well [28]. The model has been customized for Kalimpong using the methodology as described in Martelloni et al. [28]. The historical database of landslides and rainfall of Kalimpong was used to develop a customized model. The daily rainfall data were cumulated at 'n' days taking an 'n' day broad window which can shift at a time interval of one day. The values of 'n' can vary from 1 to 365. For the daily cumulated rainfall, the cumulative distribution function was plotted and was compared with a standard function for transformation [29]. This transformation connects the cumulative rainfall (z) with the target distribution (y = a.σ) where 'σ' is the standard deviation and 'a' is a multiplication factor. The rainfall series is arranged in increasing order as:

$$z_1 < z_2 < z_3 < \cdots < z_k < \cdots < z_n$$

And a cumulative frequency of sample is defined as

$$P_k = \frac{k}{n} - \frac{0.5}{n} = G(y) \tag{4}$$

where $1 \leq k \leq n$.

The values of y on the original data z can be derived as:

$$\mathrm{G}^{-1}(\mathrm{F(z)}) \rightarrow \mathrm{G}^{-1}(\mathrm{P_k}) = \mathrm{y} \tag{5}$$

Once the transformation function is applied, the cumulative sample frequency and corresponding precipitation can be estimated for any value of y. When the procedure is repeated for all values of n, the precipitation curves or σ curves can be plotted. These curves are used in the algorithm for defining threshold condition. Based on the criticality of rainfall event, algorithm issues different levels of warning. The daily rainfall cumulates are compared with the σ curves in the algorithm [28]. For shallow landslides, the model considers short-term rainfall while the long term precipitation is used for predicting deep seated landslides. The decisional algorithm used is given in Eq. (3):

$$C_{1-3} = \left[\sum_{i=1}^{n} P(t+1-i)\right]_{n=1,2,3} \geq [S_n(\Delta)]_{n=1,2,3} \tag{6}$$

where $\Delta = $ a.σ, C_{1-3} denotes the cumulated rainfall at the time of analysis t, which is a vector and $S_n(\Delta)$ are the thresholds defined by σ curves, corresponding to Δ and n days [28]. In the case of deep seated landslides, the algorithm considers the effect of precipitation from 4 days upto 63 days [28]. The condition for crossing the threshold is given by:

$$C_{4-63} = \left[\sum_{i=1}^{n+3} P(t-2-i)\right]_{n=1,2,\ldots 60} \geq \left[S_{n+3}(\Delta)\right]_{n=1,2,\ldots 60} \tag{7}$$

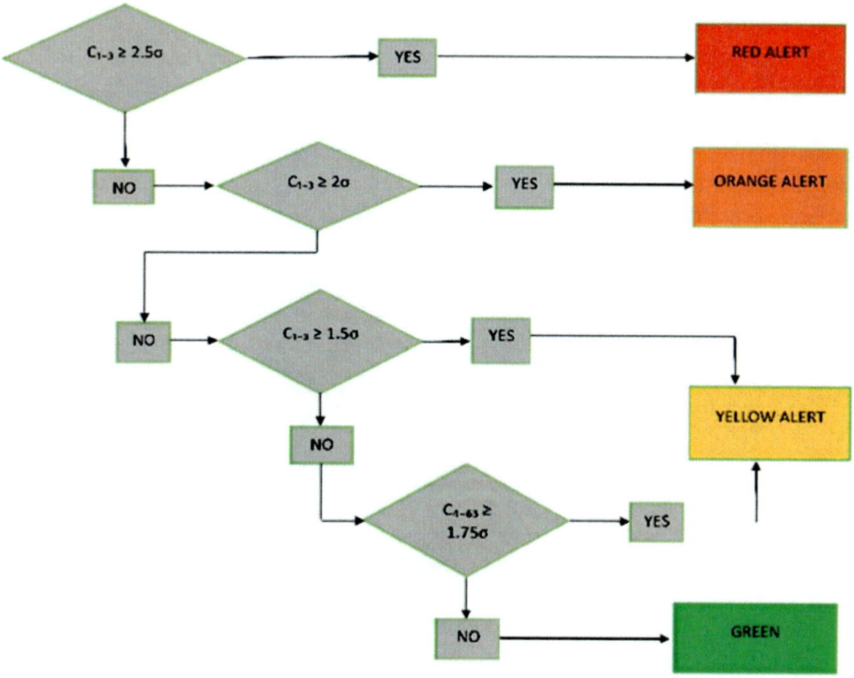

Fig. 5 Staring algorithm considered for calibration of customized SIGMA model for Kalimpong [30]

The algorithm considered for starting the calibration of the customized SIGMA model for the study area is shown in Fig. 5.

Using the procedure of optimization, false alarms are reduced, and the starting thresholds were modified. The earlier thresholds of 1.5, 1.75 and 2 were optimized to 1.65, 1.95 and 2.05 respectively.

The red alert is the most critical case where action must be taken. Orange alerts calls of preparedness and yellow alert calls for attention. On days of green alert, the region is safe any possible threats of landslides.

3.3 *Probabilistic Approach Using Soil Wetness and Rainfall Severity (RS Threshold)*

The posterior probability of initiation of landslides based on soil wetness and severity of rainfall has been done using a Bayesian analysis. It considers the possible effect of both these factors on the occurrence of landslides. For defining a hydro-meteorological threshold, the antecedent moisture content was used along with rainfall severity for defining the thresholds. A hydrological model called SHETRAN has been used for the estimation of soil moisture [22] and the rainfall severity is determined using the empirical ED thresholds using the Frequentist approach. Instead of using the automatic algorithm, all events were considered for the analysis irrespective of the relative positions of landslides and rain gauge. The soil moisture values simulated using SHETRAN were normalized using a scale from 0 to 1 for convenient classification. The normalized values scaled from 0 to 1 are known as soil wetness. The value indicates the wetness of insitu soil. Multiple levels of ED thresholds were defined using different exceedance probabilities and were classified into six categories. Thus, based on the multiple exceedance probabilities of the defined ED threshold (T_{min}, T_5, T_{10}, T_{20} and T_{50}) six categories are defined. Thus, the two-dimensional Bayesian plot contains 30 cells based on rainfall severity and soil wetness.

The Bayes theorem for two-dimensional case can be expressed as:

$$P(A|B,C) = \frac{P(B,C|A) * P(A)}{P(B,C)} \tag{8}$$

A is defined as the occurrence of at least one landslide in Kalimpong. B indicates the soil wetness value and C denotes the severity of rainfall. ‘B, C’ indicates the simultaneous occurrence of any combination of rainfall severity and soil wetness. In short, it defines the condition of each cell (5 × 6). The remaining terms in the equation can be defined as:

P(B, C|A) is defined as the probability that a certain ‘B, C’ condition occurs along with the occurrence of landslides.

P(A) is called the prior probability that a landslide occurs, regardless of B and C.

P(B, C) denotes the marginal probability that a certain cell condition occurs, irrespective of the occurrence of landslides.

P(A|B, C) is called the conditional probability that a landslide occurs when ‘B, C’ condition happens.

All these probabilities are estimated in the terms of relative frequencies.

$$P(A) \approx \frac{N_A}{N_R} \tag{9}$$

$$P(B,C) \approx \frac{N_{B,C}}{N_R} \tag{10}$$

$$P(B,C|A) \approx \frac{N_{(B,C|A)}}{N_A} \tag{11}$$

where

N_A = The total landslide events

N_R = The total rainfall events

$N_{B,C}$ = The events with the cell condition ‘B, C’

$N_{(B,C|A)}$ = The rainfall events that resulted in landslides with a cell condition ‘B, C’.

Table 2 Catchment properties calibrated using SHETRAN

Parameters	Calibrated value
Canopy storage	5 mm
Maximum rooting depth	1.6 m
Saturated hydraulic conductivity	1.14 m/day
Saturated water content	0.40
Strickler overland flow coefficient	0.50 $m^{1/3}s^{-1}$
AE/PE at field capacity	1
vanGenuchten-n	1.17
vanGenuchten-alpha	0.03 cm^{-1}
Residual water content	0.08
Leaf area index	1

Soil Moisture Estimation

The rainfall data from 2010 to 2016 was used as an input to SHETRAN model, along with the soil properties and digital elevation model. First the tool determines phreatic line based on precipitation and catchment properties. The moisture content values were then calculated using the developed phreatic line for saturated and unsaturated zones. The zone below phreatic line is always fully saturated and the moisture content varies for the unsaturated zone above phreatic line. The hydraulic parameters for modelling the water flow were calculated using van Genuchten factors in SHETRAN. Owing to the limitations induced by the assumptions taken for modelling, the outputs were calibrated for reliable outputs. The properties of catchment were modified to get an optimum value of R^2 as 0.84. The calibrated parameters are mentioned in Table 2.

The obtained results are similar to those obtained by those predicted using the particle size distribution of soil in ROSETTA Lite module [31] for Kalimpong [32].

Probabilistic Thresholds

The cell conditions for the definition of probabilistic thresholds are defined based on the rainfall severity and soil wetness classifications. There are cases when probability of occurrence of landslides is zero, as $P(B, C|A) or P(A)$ is becoming zero. The probabilistic RS thresholds defined for Kalimpong are plotted in Fig. 6.

The thresholds prove that when the soil wetness is less, very severe rainfalls are required to trigger landslides in the region and when the soil is wet, even less severe rainfalls can trigger landslides.

3.4 *Quantitative Comparison*

As stated by Lagomarsino et al. [33], the validation of a model can be done in complete form by comparing the statistical attributes of a model with those for other models

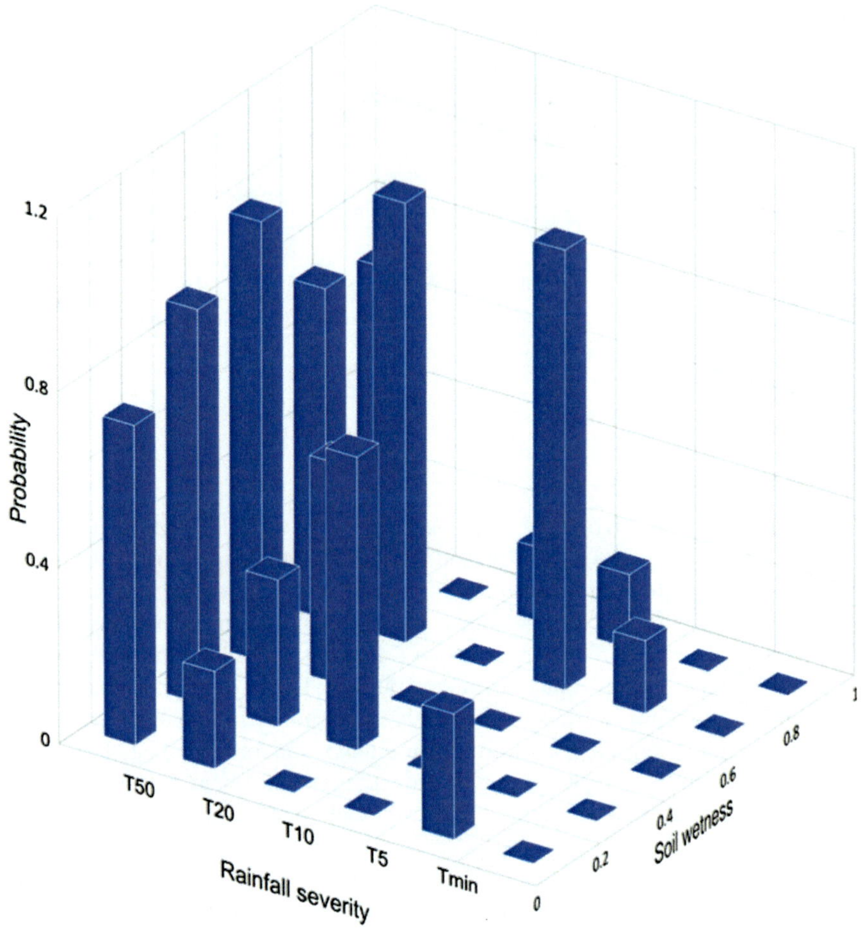

Fig. 6 Probabilistic RS thresholds for Kalimpong

defined for a region. Multiple rainfall thresholds for landslides in Kalimpong region are already defined [14, 20, 22, 30]. Therefore, all the approaches were statistically compared to find out the best suited model for Kalimpong for efficient prediction of landslides. The comparison has been done using a confusion matrix (Fig. 7). The correctly predicted landslides are true positives, missed alarms are false negatives, false alarms are false positives and the correctly predicted non-landslides are true negatives. Since there were no major landslides reported in Kalimpong, the real time field monitoring data from the tilt sensors installed in Chibo has been used for threshold validation (Table 3). The details of monitoring system and data collection is explained in Sect. 4.

The validation results show that SIGMA performs much better than the other models. The main reason is probably because slow movements occurred in 2017,

Fig. 7 Confusion matrix [30]

Table 3 Quantitative comparison of rainfall thresholds for Kalimpong

Statistical attributes	ID threshold [14]	ED threshold [20]	Antecedent rainfall threshold [14]	SIGMA model [30]	Probabilistic RS [22]
a = True positives	1	1	7	7	2
b = False positives	45	22	84	22	17
c = False negatives	6	6	0	0	5
d = True negatives	313	336	274	336	341
Efficiency = (a + d)/(a + b + c + d)	0.86	0.92	0.77	0.94	0.94
Sensitivity = a/(a + c)	0.14	0.14	1	1	0.29
Specificity = d/(b + d)	0.87	0.94	0.77	0.94	0.96
Likelihood ratio = Sensitivity/(1 – Specificity)	1.14	2.32	4.26	16.27	6.02

and a technique which uses both short term and long-term effect of landslides is better when compared to the other thresholds. The antecedent thresholds are also found to be highly sensitive, but the large number of false alarms makes it unsuitable for use in LEWS. Even though the likelihood ratio of probabilistic thresholds is better than that of antecedent thresholds, the number of missed alarms is the disadvantage of this model. However before having a threshold model more effective to use in an LEWS for Kalimpong, a validation using much larger dataset has to be used.

4 Field Monitoring

Field monitoring is an essential part of LEWS, to get real time information regarding the site conditions. Acquisition of real time and reliable data is a key parameter for risk management measures. The short-term behavior of movements may end up in drastic failures and hence to predict such incidents in advance, it is essential to perform real time monitoring. Two approaches are generally followed in recent times for real time monitoring, (a) Remote sensing-based monitoring and (b) geotechnical monitoring (inclinometers, extensometers, tiltmeters etc.). The evolution of both these methods was from the traditional site investigations conducted in regular intervals. Along railway lines and roads in hilly regions, scheduled inspections are carried out to find any abnormalities in slopes. Such investigations were helpful in identifying slow changes, but not useful to understand short term variations and predict future incidents.

4.1 Geotechnical Monitoring

Geotechnical monitoring of slopes measures the engineering behavior soil continuously. The parameters considered are the deformation (extensometers, tiltmeters etc.), moisture content (moisture content sensors), matric suction (tensiometers), scouring (scouring sensors) etc. The instruments for measuring deformation are being widely used for detection of landslide events while moisture content and matric suction are the internal factors which lead to the triggering of rainfall induced landslides and scouring can be used for real time monitoring of fast flowing rapid mass, like debris flows. Extensometers are used for in situ ground based measurements of the distance between moving and stable masses [34, 35]. The instrument demands the knowledge of probable failure planes for the installation. It requires skilled and expert judgement on possible failure, which is difficult to determine in the case of rapid movements. Uchimura et al. [36] proposed an easy alternative for this, by using tiltmeters. The method has been validated using laboratory and field tests in Japan and later in China [37–40].

The method uses tilt sensors and volumetric moisture content sensors in the slope surface and measures the rotation angle and moisture content. The wireless sensor unit consists of Micro Electro Mechanical Systems (MEMS) sensor, embedded in the unstable surface layer of slopes. The compartment includes a sensor unit inside it with a control unit, a MEMS tilt sensor and a volumetric moisture content sensor. Therefore, the system is suitable to identify the initial stages of surface failure. The abscissa of the tilt sensors is placed in the direction of the slope while the ordinate being orthogonal to it. Thus, the tilting in directions parallel and perpendicular to the slope can be recorded. Uchimura et al. [36] developed surface tilt sensor which is equipped with a MEMS tilt sensor and a volumetric water content sensor. The movement of the slope in about 0.02° can be detected using the sensors. However, the measuring precision may differ depending on weather condition and environment. The cost of each sensor unit is approximately 1000 US$. This method was adopted for Kalimpong town, for real time slope monitoring.

Real Time Field Monitoring for Kalimpong Using MEMS Tilt Sensors

In Kalimpong town, the small region of Chibo has a long history of landslides. The hills is suffering from continuous sinking during every rains [41]. This is a major transportation channel of Kalimpong, connecting the town to the National Highway. The lower order streams (jhoras) have a significant role in the downhill erosion and sinking of uphill. To study the slope movements in detail, six tilt sensor units were installed in Chibo, near two major jhoras of Chibo (Fig. 8). Sensors 1, 2 and 3 placed near Pyarieni Jhora; sensors 4 and 6 near OC jhora and sensor 5 is placed towards the central portion of Chibo, close to OC jhora. These were the critical locations identified during field investigations and are being monitored since July 2017.

Each unit consists of a control unit, volumetric moisture content sensor and a MEMS tilt sensor. The readings are recorded at every 10 min and are transferred to data loggers, using a wireless module. Four alkaline batteries of C size are used

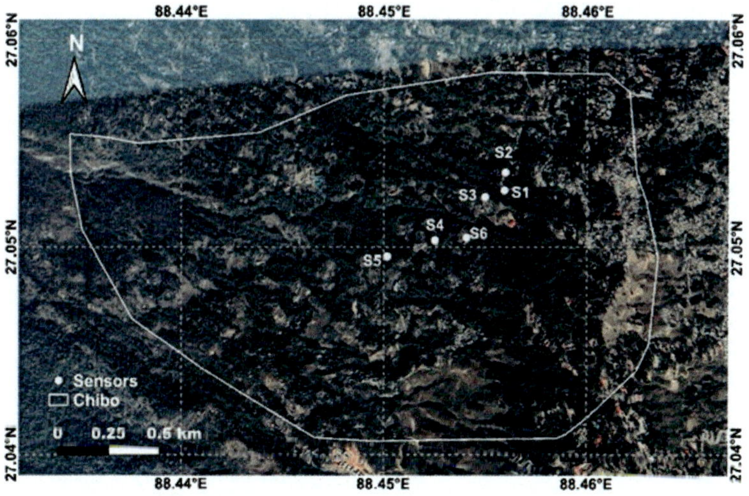

Fig. 8 Location of tilt sensors in Chibo [42]

for power supply in the sensors. Such batteries have long life and works for more than a year in field conditions. The volumetric water content measures the dielectric constant of the soil at shallow depths and calculates the volumetric water content. The slope stability of the soil depends upon the pore pressure. Considering the practical difficulties in the direct measurement of pore water pressure, volumetric moisture content is measured to understand the presence or absence of moisture in soil. The reading given by the moisture content sensor is measured only at one point, but the tilt sensor measures the behavior of soil mass it. Tilting angles in both parallel and perpendicular directions to the direction of slope are being measured using MEMS sensors.

The surface tilt sensor (Fig. 9) measures the tilt angle of the steel rod shown in Fig. 9 installed to into the unstable soil layer at a depth of 1 m or more. The sensor is placed below the ground and is connected to a wireless transmission system via the steel rod. The relative movement occurs in between the top and bottom portions of the sensor unit. When the unstable layer is thin, the average shear deformation of the surface layer of slope is detected by the sensors [37].

The recorded data from all six sensors are plotted in Fig. 10. After three monsoons of monitoring, it was observed that maximum variations occurred near sensors 2 and 3, near pyarieni jhora [41]. Sensor 2 has recorded movement during al the three monsoon seasons and hence selected for detailed analysis. Since there were no landslides recorded during the study period, the determination of threshold values for the initiation of landslides is difficult to determine.

The monitoring of slopes for continuous monsoons helps in identifying the effect of rainfall on slope stability. The most variation in tilting angles of the sensors were observed during monsoon seasons. The tilting rates are found to be proportional to the rainfall values. The details of tilting angles and volumetric moisture content data as obtained from the sensors are plotted in Fig. 10. 90.78% of the total rainfall is contributed by the monsoon season.

From Fig. 11, the variations in tilting rates with antecedent rainfall, daily rainfall, and moisture content can be understood. The most critical parameter is the antecedent rainfall, not the daily rainfall. Also, to predict the slow movements, the rainfall threshold should consider the effect of both short term and long-term rainfall. Hence, the study proves the importance of using a statistical threshold that considers the effect of both short-term and long-term precipitation as the first line of early warning [30]. Such thresholds can also be improved conceptually by incorporating hydrological parameters [21, 22, 43] and tilt angle values. The total rainfall received in the study area during the 2018 monsoon period was the least when compared to the other two monsoons.

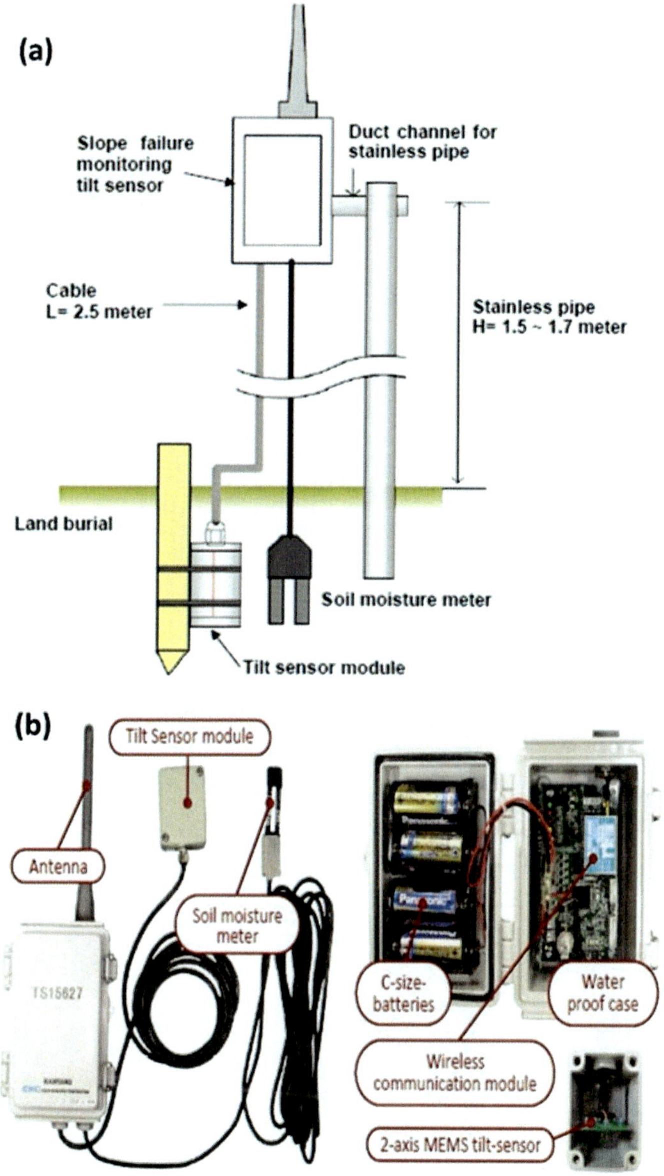

Fig. 9 **a** Schematic arrangement of the sensor unit, **b** Components of a tilt sensor [42]

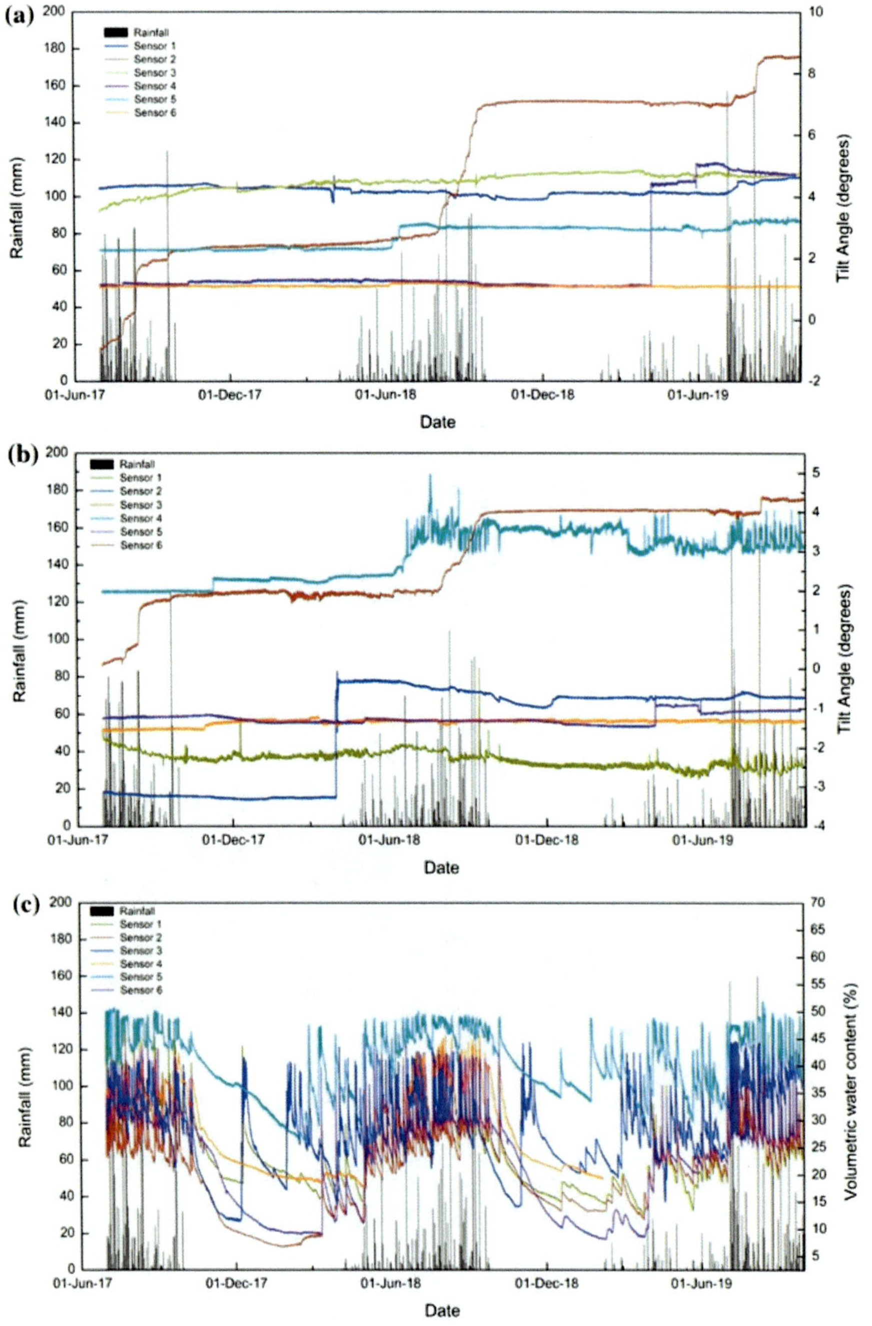

Fig. 10 Data from tilt sensors installed in Chibo **a** tilting angle (parallel to the slope), **b** tilting angle (perpendicular to the slope), and **c** volumetric moisture content

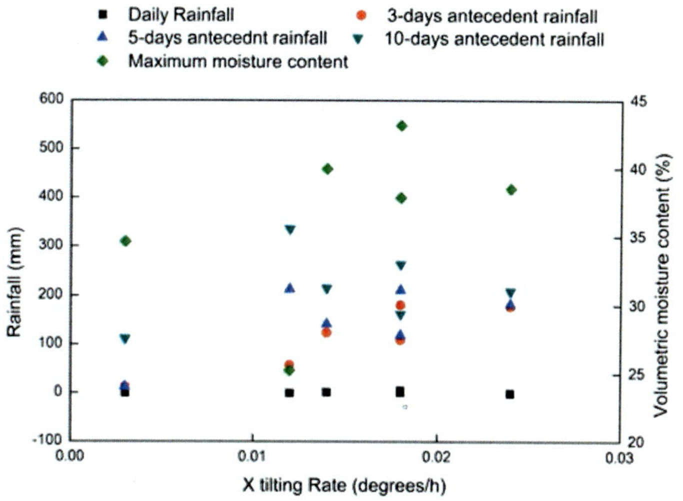

Fig. 11 Summary of readings from sensor 2 during displacement periods

5 Conclusions

The study deals with two different aspects of EWS for rainfall induced landslides, development of thresholds and field monitoring, considering the theoretical and practical aspects. The different rainfall thresholds defined for the region and the field monitoring system installed are discussed in detail. This is the first-time approach in India where multiple thresholds are defined for a study area and are validated using real time field monitoring data. It was found that the algorithm based SIGMA model predicts the landslides in Kalimpong with higher accuracy when compared to all other models. Choosing the best suited method among the threshold models is a site dependent choice. It is not mandatory that a model should work well in all hydro-geological and climatic conditions. The choice of the best suited method should be based on quantitative comparison. Hence this approach can be adopted for regions affected by rainfall induced landslides, to find the best suited landslide forecasting model.

Installation of field monitoring system has two objectives: (a) obtain reliable data in real time regarding the in situ condition, (b) validation of rainfall thresholds derived for the region. The reliability of the data has been verified with by using field observations and the same has been used for the validation of rainfall thresholds. The best suited model can be integrated with the field monitoring results to avoid false alarms and can constitute a regional LEWS.

References

1. USGS national landslide hazards mitigation strategy—a framework for loss reduction (2003)
2. Anderson MG, Holcombe E, Blake JR, Ghesquire F, Holm-nielsen N, Fisseha T (2011) Reducing landslide risk in communities: evidence from the Eastern Caribbean. Appl Geogr 31:590–599
3. Maskrey A (2011) Revisiting community-based disaster risk management. Environ Hazards 10:42–52
4. Segoni S, Rosi A, Fanti R, Gallucci A, Monni A, Casagli N (2018) A regional-scale landslide warning system based on 20 years of operational experience. Water (Switzerland) 10:1–17
5. Abraham MT, Satyam N, Rosi A, Pradhan B, Segoni S (2020) The selection of rain gauges and rainfall parameters in estimating intensity-duration thresholds for landslide occurrence: case study from Wayanad (India). Water 12:1000
6. Abraham MT, Pothuraju D, Satyam N (2019) Rainfall thresholds for prediction of landslides in Idukki, India: an empirical approach. Water 11:2113
7. Abraham MT, Satyam N, Rosi A (2020) Empirical rainfall thresholds for occurrence of landslides in Wayanad, India. In: Proceedings of the EGU general assembly 2020
8. CartoDEM National Remote Sensing Centre, Hyderabad, Department of Space, Government of India. https://bhuvan-app3.nrsc.gov.in/data/download/index.php. Accessed 20 Aug 2019
9. Chakraborty I, Ghosh S, Bhattacharya D, Bora A (2011) Earthquake induced landslides in the Sikkim-Darjeeling Himalayas—an aftermath of the 18th Sikkim earthquake 2011:1–8
10. Mukherjee A, Mitra A (2001) Geotechnical study of mass movements along the Kalimpong approach road in the Eastern Himalayas. Indian J Geol 73:271–279
11. Sumantra S (2016) Causes of landslides in Darjeeling Himalayas during June–July, 2015. J Geogr Nat Disasters 6
12. Save the hills blog. http://savethehills.blogspot.com/. Accessed 3 Dec 2019
13. Caine N (1980) The rainfall intensity-duration control of shallow landslides and debris flows: an update. Geogr Ann Ser A Phys Geogr 62:1–2, 23–27
14. Dikshit A, Satyam DN (2018) Estimation of rainfall thresholds for landslide occurrences in Kalimpong, India. Innov Infrastruct Solut 3:24
15. Brunetti MT, Peruccacci S, Rossi M, Luciani S, Valigi D, Guzzetti F (2010) Rainfall thresholds for the possible occurrence of landslides in Italy. Nat Hazards Earth Syst Sci 10:447–458
16. Guzzetti F, Peruccacci S, Rossi M, Stark CP (2007) Rainfall thresholds for the initiation of landslides in central and southern Europe. Meteorol Atmos Phys 98:239–267
17. Cannon S, Ellen S (1985) Rainfall conditions for abundant debris avalanches, San Francisco Bay region, California. Calif Geol 38:267–272
18. Peruccacci S, Brunetti MT, Luciani S, Vennari C, Guzzetti F (2012) Lithological and seasonal control on rainfall thresholds for the possible initiation of landslides in central Italy. Geomorphology 139–140:79–90
19. Melillo M, Brunetti MT, Peruccacci S, Gariano SL, Guzzetti F (2014) An algorithm for the objective reconstruction of rainfall events responsible for landslides. Landslide dynamics: ISDR-ICL landslide interactive teaching tools. Volume 1: Fundamentals, mapping and monitoring, vol 12, pp 311–320
20. Teja TS, Dikshit A, Satyam N (2019) Determination of rainfall thresholds for landslide prediction using an algorithm-based approach: case study in the Darjeeling Himalayas, India. Geosciences 9:302
21. Zhao B, Dai Q, Han D, Dai H, Mao J, Zhuo L (2019) Probabilistic thresholds for landslides warning by integrating soil moisture conditions with rainfall thresholds. J Hydrol 574:276–287
22. Abraham MT, Satyam N, Pradhan B, Alamri AM (2020) Forecasting of landslides using rainfall severity and soil wetness: a probabilistic approach for Darjeeling Himalayas. Water 12:804
23. Lumb P (1975) Slope failure in Hong Kong. Q J Eng Geol 8:31–65
24. Giannecchini R (2005) Rainfall triggering soil slips in the southern Apuan Alps (Tuscany, Italy). Adv Geosci 2:21–24

25. Melillo M, Brunetti MT, Peruccacci S, Gariano SL, Roccati A, Guzzetti F (2018) A tool for the automatic calculation of rainfall thresholds for landslide occurrence. Environ Model Softw 105:230–243
26. Gariano SL, Sarkar R, Dikshit A, Dorji K, Brunetti MT, Peruccacci S, Melillo M (2018) Automatic calculation of rainfall thresholds for landslide occurrence in Chukha Dzongkhag, Bhutan. Bull Eng Geol Environ 1–8
27. Dikshit A, Satyam N (2017) Rainfall thresholds for the prediction of landslides using empirical methods in Kalimpong, Darjeeling, India. In: Proceedings of the JTC1 workshop on advances in landslide understanding; Barcelona, Spain
28. Martelloni G, Segoni S, Fanti R, Catani F (2012) Rainfall thresholds for the forecasting of landslide occurrence at regional scale. Landslides 9:485–495
29. Goovaerts P (1997) Geostatistics for natural resources evaluation. Oxford University Press, Oxford
30. Abraham MT, Satyam N, Kushal S, Rosi A, Pradhan B, Segoni S (2020) Rainfall threshold estimation and landslide forecasting for Kalimpong, India using SIGMA model. Water 12:1195
31. Schaap MG, Leij FJ, Genuchten MT Van (2001) ROSETTA: a computer program for estimating soil hydraulic parameters with hierarchical pedotransfer functions. J Hydrol 251:163–176
32. Dikshit A, Satyam N, Pradhan B (2019) Estimation of rainfall-induced landslides using the TRIGRS model. Earth Syst, Environ
33. Lagomarsino D, Segoni S, Rosi A, Rossi G, Battistini A, Catani F, Casagli N (2015) Quantitative comparison between two different methodologies to define rainfall thresholds for landslide forecasting. Nat Hazards Earth Syst Sci 15:2413–2423
34. Intrieri E, Gigli G, Mugnai F, Fanti R, Casagli N (2012) Design and implementation of a landslide early warning system data acquisition. Eng Geol 147–148:124–136
35. Angeli MG, Pasuto A, Silvano S (2000) A critical review of landslide monitoring experiences. Eng Geol 55:133–147
36. Uchimura T, Towhata I, Lan Anh TT, Fukuda J, Bautista CJB, Wang L, Seko I, Uchida T, Matsuoka A, Ito Y et al (2010) Simple monitoring method for precaution of landslides watching tilting and water contents on slopes surface. Landslides 7:351–357
37. Uchimura T, Towhata I, Wang L, Nishie S, Yamaguchi H, Seko I, Qiao J (2015) Precaution and early warning of surface failure of slopes using tilt sensors. Soils Found 55:1086–1099
38. Yang Z, Cai H, Shao W, Huang D, Uchimura T, Lei X, Tian H, Qiao J (2018) Clarifying the hydrological mechanisms and thresholds for rainfall-induced landslide: in situ monitoring of big data to unsaturated slope stability analysis. Bull Eng Geol Environ 78:2139–2150
39. Towhata I, Uchimura T, Seko I, Wang L (2015) Monitoring of unstable slopes by MEMS tilting sensors and its application to early warning. IOP Conf Ser Earth Environ Sci 26:012049
40. Xie J, Uchimura T, Wang G, Shen Q, Maqsood Z, Xie C, Liu J, Lei IW, Tao IS, Chen IP et al (2019) A new prediction method for the occurrence of landslides based on the time history of tilting of the slope surface. Landslides
41. Dikshit A, Satyam DN, Towhata I (2018) Early warning system using tilt sensors in Chibo, Kalimpong, Darjeeling Himalayas, India. Nat Hazards 94:727–741
42. Abraham MT, Satyam N, Pradhan B, Alamri AM (2020) IoT-based geotechnical monitoring of unstable slopes for landslide early warning in the Darjeeling Himalayas. Sensors 20:2611
43. Segoni S, Rosi A, Lagomarsino D, Fanti R, Casagli N (2018) Brief communication: using averaged soil moisture estimates to improve the performances of a regional-scale landslide early warning system. Nat Hazards Earth Syst Sci 18:807–812

India and China Scientific Collaborations at Grass-Root Level: A New Era

Da Hsuan Feng and Ankit Garg

Abstract COVID-19 marks one of the most challenging year in human history with unprecedented globalization challenges. However, there is a silver lining in the midst of such challenges in that it invigorated intensive and unlimited academic cooperation at grass-roots level between the world's two most ancient civilizations, India and China. It was a year of 1st Indo-China Research Webinar Series featuring over 5000 participants, 16 technical sessions and 10 Special Guests. Under normal circumstances, this clearly would not be possible. After all, generally, it would have required at least a year of detailed planning coupled with significant funding. The success of the series stood on the foundation of the past several years of scientific collaborations. This chapter describes this scientific cooperation in terms of youth exchange (2-way) between outstanding universities from both countries. In addition, the impact of such cooperation in terms of Joint International Research Awards has also been presented. Still, what is presented here is still only the tip of iceberg of the impact of scientific cooperation between the youth of both countries in past few years. Undoubtedly, a great deal still needs to be done to further deepen relations in people to people exchange at the scientific level between both countries, especially in seeking solutions to scientific issues of mutual interest, especially in the global existential challenges of climate change and the economic future of sustainable infrastructure development for humanity.

Keywords Indo-China webinar · Academic cooperation · Grass-roots level · International research award · Youth exchange

D. H. Feng
University of Texas at Dallas, Richardson, USA

China Silk Research iValley Institute, Hong Kong, China

Chairman of the International Advisory Board, Hainan University, Haikou, China
e-mail: dahsuan@gmail.com

A. Garg (✉)
Guangdong Engineering Center for Structure Safety and Health Monitoring, Shantou University, Shantou, China
e-mail: ankit@stu.edu.cn

A. Garg et al. (eds.), *Proceedings of the 1st Indo-China Research Series in Geotechnical and Geoenvironmental Engineering*, Lecture Notes in Civil Engineering 123,
https://doi.org/10.1007/978-981-33-4324-5_12

1 Introduction

The year 2020 marks the "70th anniversary of diplomatic exchange" between India and China, two of the world's most ancient civilizations. The mutual diplomatic recognition on January 1st, 1950, is especially remarkable since India is the first non-Communist nation to do so.

Of course, year 2020 will also be recorded in human history due to the emergence of COVID-19 pandemic, which has created unprecedented and severe global health and geopolitical challenges.

Quite unexpected is that the health challenge has become the driving force in the enhancement of a global efforts in the containment of this pandemic as well as the formation of new chains of academic cooperation between various countries. To this end, young scholars from India and China, who have already been pro-active in collaboration for a number of years, leveraged this opportunity to organize the "1st Indo-China Webinar Research Series," which lasted from the 8th to the 19th of May, 2020. Standing on the 5G technology platform through ZOOM, the series provided live-streamed lectures to and interactions for more than 800–1000 participants (each session) worldwide.

Unlike "normal" conferences of this magnitude which would require many months of meticulous and arduous planning, this conference was literally "organic" in its conception and almost spontaneously organized and implemented by the enormous enthusiasm of many young faculty members in both countries.

1.1 1st Indo-China Research Webinar Series 2020

This series was collaboratively organized by.

1. Dr. Ankit Garg, Associate Professor, Shantou University, China,
2. Prof. (Dr.) Chandresh H. Solanki, Professor, SVNIT, Surat & Indian Geotechnical Society Surat Chapter, Gujarat, India,
3. Dr. Chandra Bogireddy, Assistant Professor, Vardhaman College of Engineering, Hyderabad, India and
4. Dr. Junwei Liu, Vice Dean & Associate Professor, Qingdao University of Technology, China.

This Indo-China research webinar had 16 technical sessions, with 8 Speakers in each coming from universities in China and India. The total number of participants in all sessions were more than 5000, participating electronically primarily from China and India.

It is worth underscoring that 11 Distinguished Guests were invited to present to the audience their motivational words to promote such cooperation in the future. They are.

1. Prof. G. L. Sivakumar Babu, President, Indian Geotechnical Society & Professor IISc Bangalore,
2. Prof. S.R. Gandhi, Director, S.V. NIT Surat, India,
3. Prof. Gautam Biswas, Former Director, Indian Institute of Technology (IIT) Guwahati, Fellow of ASME,
4. Prof. Da Hsuan Feng, Fellow American Physical Society, USA, Chairman of Advisory Board, Hainan University, Former Vice President for Research, University of Texas, USA,
5. Prof. S. K. Das, Director, IIT Ropar,
6. Prof. D. N. Singh, Institute Chair Professor, IIT Bombay & Editor in Chief, Environmental Geotechnics,
7. Prof. Chandan Ghosh, IIT Jammu, NIDM, Ministry of Home affairs, India
8. Prof. Askar Zhussupbekov, Past Vice President of ISSMGE for Asia & Director of the Geotechnical Institute.
9. Prof. Neelamani, Senior Research Scientist, Kuwait Institute of Scientific Research,
10. Dr. Jaykumar Shukla, Principal Engineer, Geo Dynamics, India and
11. Mr. Subba Reddy Nelluru (UK).

The forum is also supported by Dr. S. Sai Satyanarayana Reddy, Principal, VCE and Dr. K. Mallikharjuna Babu, Director and CEO, Vardhaman College of Engineering Hyderabad.

The success of the event was defined by the ambiance of warm and close interactions and networking among the young scholars, many were (electronically) meeting for the first time. Moreover, the event was also extensively covered by the International Society of Soil Mechanics and Geotechnical Engineering [1].

Perhaps what is most remarkable is that this research webinar series was de facto a self-discovery process. In fact, under "normal" circumstance, this would not be possible owing to the fact that not until recently, the digital technology was lacking in robustness for large scale simultaneous conferencing, and the requirement of excessive budget. Having had this experience under one's belt, the organizers of the series are already now making plans to organize such conferences on an annual basis. Many participants mentioned that the certificates issued to those who had greater than 75% attendance have left an indelible mark of this great event in their minds.

It is not just the voice of scientists from China and India but also from other countries (USA, Sri Lanka), that is vouching for enhanced collaboration in environmental research between two countries as evident from article "*China and India: Toward a sustainable world*" [2, 3]. The article suggests "*Science diplomacy*". It also suggests opening joint research centers related to environment especially at Himalayas, which are amongst the highest peaks in world. Such scientific cooperation at Himalayas can enhance trust, understanding and lead to a more sustainable world in future. The article provides a unique idea of converting Himalayas, from barely rocks into a mountain of scientific cooperation between countries.

1.2 Research Collaborative Schemes from Indian and Chinese Sides

Although the COVID-19 was the incentive for this Webinar forum, it needs to be pointed out that within the last few years, Indians and Chinese scientific and technological collaboration was already manifesting significant rise. As evident from the article published by News in Asia [4], there has been a strong increase in Indian students currently pursuing education and research opportunities in China. Indeed, according to an article published in Times of India, in comparison to the more traditional destinations such as the United Kingdom, United States, Australia and Canada, China has emerged to become another favorite destination for Indian students pursuing advanced studies [5].

Just recently, India and China have managed to establish various unprecedented collaborations, through a new program called SPARC or "Scheme for Promotion of Academic and Research Collaboration" [6]. This program was initiated by the vast and powerful IITs (Indian Institute of Technology) system and other top Indian universities. Under the umbrella of the IIT system, there are 23 outstanding technological universities [7]. The financial support of the program originates from India's Ministry of Education (formerly known as Ministry of Human Resource Development). The primary mission of this program is to be the conduit to establish strong, meaningful and sustainable international cooperation between top Indian universities with their counterparts in some of the 30 top-notch Chinese universities. It should be mentioned that the program also includes collaborations of the said institutions in India with other top universities worldwide, including MIT, Stanford etc.

In 2018, a number of Chinese universities have been awarded joint projects [8]. This year (2019), SPARC has already made the call-for-proposals with results awaiting.

With the above information as background, as far as scientific and technological platforms are concerned, contrary to what has been reported in the media, India has in fact and to a great extent deepened the collaboration with China. The program SPARC which was initiated in 2018 has become one of the flagships of a variety of such Indo-China programs.

It is worth underscoring that SPARC is not a hollow program, but one which puts "money where its mouth is." For example, for a faculty who meets the requirement threshold to spend one academic year in an India university to conduct collaborative research, he/she shall receive a stipend of up to $15,000 USD per month.

Other programs such as VAJRA (or Visiting Advanced Joint Research,) GIAN (or Global Initiative of Academic Networks) and an assorted University-based schemes initiated by individual IIT campuses to invite foreign faculty for visiting their respective campuses have also created a positive atmosphere to increase visits by Chinese faculty members to India, either as Adjunct or Visiting Faculty in various IITs.

For example, Prof Zhou Hongfu from the South China University of Technology in Guangzhou has joined IIT Delhi as a Visiting Professor [9]. In November 2019, Dr. Changqing Chen from Tsinghua University, Beijing will deliver a speech at the workshop on "Mechanics of Mechanical Metamaterials" at IIT Madras under GIAN sponsorship [10]. Last but not least, Huazhong University of Science and Technology in Wuhan and Shanghai Jiaotong University have in past years recruited a number of young Indian faculty members namely, Dr. Akhil Garg and Dr. Kolan Madhav Reddy, respectively.

There is also program known as "海外名师" (Overseas outstaniding teachers project), which is meant to invite eminent faculties from all over world to spend long or short time in China. Recently, in August 2020, Dr. Neelima Satyam, who is currently Associate Professor and Head of Department of Civil Engineering, IIT Indore) have successfully been nominated by such a program to visit Shantou University to establish research cooperation. She is currently an executive member of the Indian Geotechnical Society. She wanted to establish collaboration with the faculty members in Shantou University in landslide monitoring and also bio-remediation technology for treatment of soil treatment.

1.3 Impact of Youth Research Exchange between Indian and Chinese Universities: A New Beginning

One should also underscore that in the past several decades, the quality of research in many Chinese universities have risen to the level where they become one of the global centers to attract fresh Indian Ph.D.s to pursue postdoctoral trainings. For example, Mr. Vinay Sharma, who is currently enrolled in a postdoctoral program at Zhejiang University, suggests that for Indians students and academicians, China actually is the closest next superpower destination in terms of technological innovation. Such a closeness in distance and economical air transport provides Indian students and young scientists an added incentive to remain closely in touch with their family.

It was evident during the past three years from number of students (at both undergraduate and postgraduate level) from elite institutions such as IIT Guwahati, IIT Jodhpur, IIT Ropar, IIT Bhubaneswar [11] and Mahindra École Centrale, Hyderabad campus visited Shantou University as well as other universities such as Guangxi University and Qingdao University of Technology (refer to Figs. 1, 2 and 3). These students from India were not in China merely to participate in the conventional course-based exchange. Rather they are there to work hand in hand with their Chinese counterparts in joint research projects. Such exchanges could be regarded as the new experiments to promote "research-based communications," which inculcates strong "cultural communications and bonding" among young students between the two countries. It is different from conventional classroom teaching atmosphere, where students rarely get the opportunity and the true "motivation" to join hands to understand and integrate each other's "mindset" to solve problems.

One of the major outcomes was the international recognition of joint research carried out by visiting students. The joint research conducted by Mr. Phani Gopal from Mahindra École Centrale, Hyderabad campus, Dr. Sanandam Bordoloi (IIT

Guwahati), Dr. Junjun Ni (Hong Kong University of Science and Technology, Hong Kong, China) and Mr. Weiling Cai (Shantou University) was awarded International Telford Award from the British Civil Engineers Association [12]. The work was jointly supervised under the Vice-Dean, Prof Lin Peng (Shantou University), Prof Mei Guoxiong (Changjiang Scholar (长江学者 in Chinese), Dean of College of Civil Engineering and Architecture, Guangxi University), Dr. Ankit Garg (Shantou University) and Dr. Tjalfe Poulsen (Associate Professor, Guangdong-Technion Israel Institute of Technology (GTIIT), China). International exposure of students through GTIIT, a Sino-foreign cooperative educational institution established by Shantou University (STU) and the Technion—Israel Institute of Technology (theTechnion), can also provide a new "dimension" in scientific mindset of youth.

In addition, Dr. Gangadhar Reddy (Fig. 4), who received funding from the Chinese Scholarship Council conducted collaborative research project with Prof Honghu Zhu, who is a Professor at the School of Earth and Sciences of Nanjing University (Fig. 5).

Also, the research group at Nanjing University led by Prof Shi Bin (Recipient of Distinguished Young Scholars Fund of National Natural Science Foundation of China (杰青) and First Prize of National Science and Technology Progress Award of China) is the pioneer in developing state of art optical fiber technology [13] for real-time monitoring of geo-engineering infrastructure. Dr. Reddy said he greatly values such experiences and wanted to implement this advanced technology in an infrastructure projects in India in the future.

Similarly, Mr. Himanshu Kumar from IIT Guwahati along with Mr. Rahul Balaji and Mr. Sadashiv S.G. from IIT Bhubaneswar conducted research exchange at Guangxi University under the supervision of Prof Guoxiong Mei Guoxiong (Figs. 1 and 2). Prof Guoxiong Mei is the recipient of the Youth Award of Mao Yisheng Soil mechanics and Geotechnical Engineering, 2019 (茅以升土力学及岩土工程青年奖) and also member of Teaching Supervisory Committee, Civil Engineering of colleges and universities, Ministry of Education (教育部高等学校土木类专业教学指导委员会). Visiting students (from IIT Bhubaneswar) at Guangxi University worked on design and development of novel stable carbon for enhancing water sorption capacity of sponge city infrastructure. Such international collaborative efforts and support from administration of these universities made such visits highly useful from the scientific point of view. This could have long term impact in knowledge sharing and development of joint technology for infrastructures.

It is certainly not a surprise for anyone to find an American Chinese studying in China, but it was astonishing to locate two American born Indian sisters studying MBBS program at Shantou University Medical College. From the authors brief conversation with them, it was discovered that their parents (American Indians) had in-depth knowledge of the Chinese education system and the accompanying career prospects. Not only in terms of students, but also many faculty members who are currently employed in Chinese universities, who are Asians with foreign origin. Therefore, China is not only becoming an attractive destination for students within Asia, but also to Asian Americans.

One should never forget the profound contributions of Dr. Dwarkanath Shantaram Kotnis (or known in China with his Chinese name Ke Dihua; Chinese: 柯棣华),

Fig. 1 Mr. Rahul Balaji and Mr. Sadashiv S. G. (IIT Bhubaneswar), Mr. Himanshu Kumar (IIT Guwahati), Mr. Huang Shan and Mr. (Guangxi University) with Prof Guoxiong Mei (Changjiang Scholar (长江学者) and Dean of College of Engineering, Guangxi University, China)

who was one of the five Indian physicians dispatched to China to provide medical assistances during Sino-Japanese War in 1938. In fact, in the south side of Martyr's Memorial park in Shijiazhuang city of the Northern Chinese province of Hebei, there is a statue of D. Kotnis, where people pay tributes to remember his contribution. In addition, a school is named after Dr. Kotnis as the Shijiazhuang Ke Dihua Medical Science Secondary Specialized School. Soon, there would be a bronze statue to be erected outside a medical school in north China, which will be formally unveiled in September 2020. This shows that such eternal bonding has existed even in terms of toughest times during history between scientists of both nations.

1.4 1st Research Visit of Chinese students to IIT Guwahati

In addition to above research output, another major outcome from the above discussed exchanges of Indian students was the reciprocal visit of Chinese students (Mr. Weiling Cai and Mr. Peinan Chen; Fig. 4) to IIT Guwahati campus. As far as the authors are aware of, this visit could be the first ever visit of Chinese students to IIT Guwahati

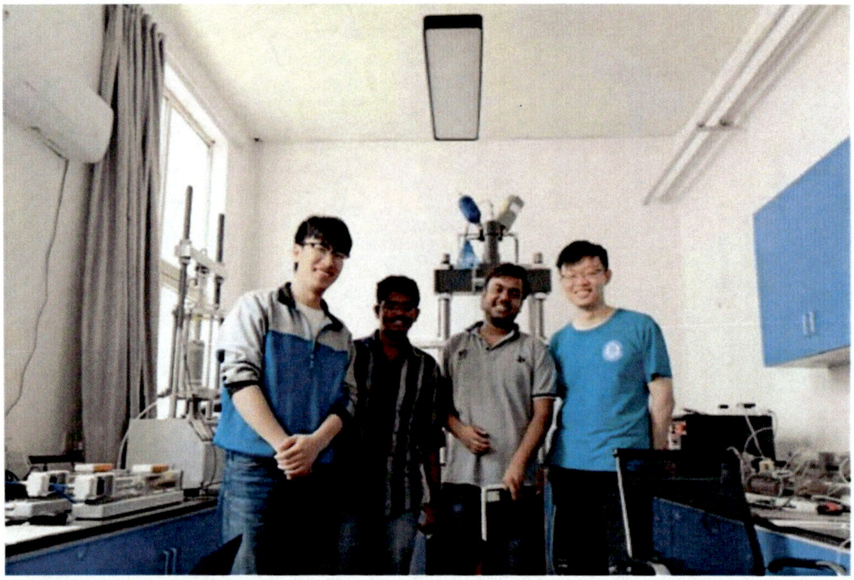

Fig. 2 Visit of IIT Bhubaneswar Students (Karthik Dutta and Abhishek Mishra) to Soil Dynamics Laboratory of Qingdao University of Technology headed by Prof Junwei Liu

campus, even though students from Japan, South Korea, Africa and middle east Universities have made regular annual visits for quite some time.

According to Mr. Weiling Cai: "It was in fact the "confidence" that was built up from previous visits of Indian students that motivated me to explore India's academics."

Unquestionably, the visit by Mr. Cai can be regarded as highly unusual for a Chinese student who comes from Guangdong province, which is one of the economically most developed provinces of China. Most students in Guangdong tend to visit developed countries such as UK, USA and Canada for the standard academic exchanges during and throughout their undergraduate careers. It was purely the cultural bonding with the visiting Indian students in Shantou University that ultimately motivated him to engage an academic visit to IIT Guwahati.

Not surprisingly, after Mr. Cai's successful visit, another student, Mr. Peinan Chen decided to follow suit to visit India.

India is a nation of vast cultures and languages. Indeed, these visits by Chinese students were highly successful not only in terms of joint research output (as mentioned above,) but also in significantly enhancing their appreciation of the diverse "Indian culture" by the Chinese students.

For example, Mr. Weiling Cai took the opportunity during his stay in India to participate in an international conference held at IIT Roorkee, which is one of the oldest campus (known as Thomason College of Civil Engineering; 1854–1947) in India established during the British Era. The visit was specifically mentioned in the admission brochure for the 2020 students. The news of his visit was also covered by Indians in China (IIC) group, which is a well-known platform to share culture and exchange between two countries. As a result of such efforts, Mr. Weiling Cai was also awarded a national scholarship and MPhil scholarship to study at Hong Kong University of Science and Technology, which is ranked amongst the top 50 universities in world.

Similarly, Mr. Liu Yun, who had jointly conducted research with Indian students was awarded prestigious national scholarship and also international scholarship for conducting Ph.D. at University of Technology, Sydney in 2020.

Dr. Vikas Pratap Singh (Ph.D. in 2018, IIT Jodhpur) who was hired by Eaton Research Laboratory in Pune (a major city located in Maharashtra province of India) is grateful for his research visit and learning he received in China. In fact, his employer even made a telephone call to his supervisor, Dr. Akhil Garg (then at Shantou University) about the ability of Dr. Vikas Pratap Singh. Table 1 summarizes student exchanges between Indian and Chinese universities as initiated by Dr. Ankit Garg and his collaborators.

1.5 *Chinese Delegation Visits to IIT Guwahati, IIT Bhubaneswar and NIT Meghalaya*

Apart from the youth research exchanges, there has been significant increase in delegation visits for scientific exchange between India and China. Some of these visits were personally covered by Dr. Ankit Garg.

It was early winter in 2018, when a research team (Fig. 7) from Shantou University visited IIT Guwahati to explore cooperation in several cutting-edge areas including green infrastructure, electric vehicles etc. During their visit, they met Prof Gautam Biswas (then Director of IIT Guwahati, Fellow of ASME) and Administration from Alumni and External Relations office (Dean, Prof Ravi Mokashi Punekar and Associate Dean, Prof Rakhi Chaturvedi (Now Dean)).

The team from Shantou University introduced to their Indian hosts the Chinese higher education system. Ideas were exchanged as to how to enhance and scale up cooperation between IIT Guwahati and Shantou University. This delegation visit resulted in exchanges of students from both sides (Figs. 3 and 6). Similarly, In January 2019, this delegation also made visits to IIT Bhubaneswar (Fig. 8) and National Institute of Technology (NIT) Meghalaya (Fig. 9) which led to exchanges of students at Guangxi University (Fig. 1), Qingdao University of Technology (Fig. 2) and Nanjing University (Fig. 5). A short workshop on "Fundamentals of Academic Writing" was conducted for improvement of scientific writing skills of Ph.D. scholars [14]. Recently, the delegation visit to IIT Indore campus has led to joint publications and also overseas famous project application of Dr. Neelima Satyam (Associate Professor and Head of Department of Civil Engineering, IIT Indore) to visit China in 2021.

Table 1 Summary of selected student exchange between Indian and Chinese Universities[a]

Name	Research visit type, organization and timeline	Current position/remarks
Dr. Sanandam Bordoloi (Former Ph.D. Scholar, IIT Guwahati, India)	Visiting Scholar, Shantou University (October 2017–January 2018)	Post-Doctoral Fellow, at Hong Kong University of Science and Technology (HKUST), Hong Kong, China [a]Co-author in International Telford Premium Prize from British Civil Engineers Association
Mr. Himanshu Kumar (Former undergraduate student, IIT Guwahati)	Visiting Scholar, Shantou University and Guangxi University (January–June, 2019)	Ph.D. at HKUST, Hong Kong, China
Mr. Suriya Ganesan (Former research Scholar, IIT Guwahati, India)	Research Associate, IIT Guwahati (May 2018–June, 2019)	Master Program at Qingdao University of Technology, China
Mr. Shubham Gaurav (Undergraduate student, IIT Guwahati, India)	Research Internship to Shantou University (May–July 2019)	–
Mr. Palash Mathur (Former undergraduate student, IIT Guwahati, India)	Research Internship to Shantou University (May–June, 2018)	–
Mr. Phani Gopal Maddibiona (Former undergraduate student, Mahindra École Centrale, India)	Research Internship to Shantou University (May–August, 2018)	Co-author in International Telford Premium Prize from British Civil Engineers Association
Mr. Weiling Cai (Former Undergraduate student, Shantou University, China)	Research exchange to IIT Guwahati (August 2019–January, 2020)	MPhil at Hong Kong University of Science and Technology, Hong Kong, China [a]Co-author in International Telford Premium Prize from British Civil Engineers Association
Mr. Peinan Chen (Former Undergraduate student, Shantou University, China)	Research exchange to IIT Guwahati (January–June 2020)	Engineer at China Construction Ltd
Dr. Vikas Pratap Singh (Former Ph.D. Scholar, IIT Jodhpur, India)	Research Internship to Shantou University (May–July, 2018)	Senior Research Engineer Eaton Research Laboratory, Pune, India
Dr. Surinder Singh (Former Ph.D. Scholar, IIT Ropar, India)	Research Internship to Shantou University (May–August, 2018)	Post-Doctoral Fellow, Université Laval, Quebec, Canada

(continued)

Table 1 (continued)

Name	Research visit type, organization and timeline	Current position/remarks
Mr. A.K. Jishnu (Undergraduate student, IIT Bhubaneswar, India)	Research Internship to Shantou University (May–July, 2019)	–
Mr. Dosetti Karthik Datta (Undergraduate student, IIT Bhubaneswar, India)	Research Internship to Qingdao University of Technology, China (May–July, 2019)	–
Mr. Aman Singhal (Undergraduate student, IIT Bhubaneswar, India)	Research Internship to Qingdao University of Technology, China (May–July, 2019)	–
Mr. Rahul Balaji (Undergraduate student, IIT Bhubaneswar, India)	Research Internship to Guangxi University, China (May–July, 2019)	–
Ms. Insha Wani (Ph.D. Student, IIT Jammu, India)	Co-Ph.D. supervision with Dr. Vinod Kushvaha, IIT Jammu	–
Mr. Sri Prasanna (Former undergraduate student, Mahindra École Centrale, India)	Research Internship to Shantou University (May–July 2018)	–
Mr. Sri Krishna (Former Master student, NUS Singapore)	Research Internship to Shantou University (May–June, 2018)	–
Mr. Chaitanya Ruhatiya (Former undergraduate student, Mahindra École Centrale, India)	Research Internship to Shantou University (May–July, 2018)	–
Ms. Sudipta Sarmah Bijoy (Ph.D. Scholar, IIT Guwahati, India)	Research Internship to Shantou University (March–April, 2018)	Ph.D. Scholar at IIT Guwahati
Mr. Raval Ratnam (Former undergraduate student, Mahindra École Centrale, India)	Research Internship to Shantou University (May–July, 2018)	–
Ms. Hima Sankari (Former undergraduate student, Mahindra École Centrale, India)	Research Internship to Shantou University (May–July, 2018)	–
Ms. Rishita Boddu (Former undergraduate student, Mahindra École Centrale, India)	Research Internship to Shantou University (May–July, 2018)	–
Dr. Ankit Goyal (Ph.D., NIT Jaipur, Chinese Scholarship Council Awardee)	Visiting Scholar, Shantou University (September 2018–July, 2019)	Post-Doctoral, University of Amsterdam, Netherlands

(continued)

Table 1 (continued)

Name	Research visit type, organization and timeline	Current position/remarks
[b]**Dr. Gangadhar Reddy** (Ph.D., IIT Bhubaneswar, Chinese Scholarship Council Awardee)	Visiting Scholar, Shantou University (September 2019–June, 2020)	–

[a,b]Selected exchange as per authors knowledge and in-depth background. Authors are themselves involved in the exchange process.

Fig. 3 Students from IITs and Mahindra Ecole Centrale with Vice Dean Prof Peng Lin (Shantou University)

2 Concluding Remarks

India and China have enjoyed millenniums of peaceful and mutually profitable co-existence. A shining example of these two countries' interaction was when Xuan Zhen (602–664 AD) in the Tang Dynasty made the long and arduous visit to India to obtain the original writings of Buddha, thus allowing Buddhism to flourish in China and other parts of East Asia. However, despite such celebrated historical events, with present geopolitics, there still exist much room to improve the cultural communications between the two countries.

To this end, in India today, there are a few truly distinguished individuals who have devoted their intellectual pursuits to create a better ambiance for both countries and, and more importantly, deeper understand of China, historically, politically, technologically and intellectually.

Fig. 4 Dr. Ankit Garg with Prof Honghu-Zhu (Nanjing University), Dr. Reddy (Visiting Scholar from India through Chinese Scholarship Council) and Prof Lei Gao (Hohai University) along with other students from Hohai University

One such individual is Professor Bali Deepak of Jawaharlal Nehru University (Fig. 10). Not only has Professor Deepak been a constant visitor to China to discuss and propose more congenial Sino-India geopolitics, he has also completed the monumental task of translating the ancient FOUR BOOKS (四书), namely, Great Learning (大学), Doctrine of the Mean (中庸), Analects (论语) and Mencius (孟子) into Hindi. The content of these four books, without a doubt, defines the most profound Chinese mindset. Their translations can and will provide for the vast Indian population a deeper and comprehensive understanding at its core the Chinese philosophy and ways and means!

Of course, the above-mentioned programs are merely some of the many examples indicating that one is observing the fledgling sea change between India and China engaging in scientific and technological collaboration. For two nations with a combined population of 2.6 billion (roughly 40% of the world's population,) this is at best only a tiny effort.

Such a collaboration should indeed be globally meaningful since China and India are countries with ancient and robust civilizations and wisdoms along with a combined 40% of the world's population. Undoubtedly, these outcomes are fundamentally the consequence of India and China placing top priorities in their respective unlimited pursue of scientific and technological innovations and excellence in nation building. We are confident that in the coming years, scholars from these two great

Fig. 5 Dr. Ankit Garg with faculties from Nanjing University. From left to right: Prof Chao-Sheng Tang, Prof Baojun Wang, Dr. Ankit Garg, Prof Bin Shi (杰青), Prof Honghu-Zhu and Dr. Qing Cheng

nations can and will continue to collaborate hand-in-hand to ensure that their collaborations will not only value-add to their nations, but that it will be a shining example of how humans can work together to overcome existential challenges.

The two authors would like to reminisce their first meeting (i.e., University of Macau; refer to Fig. 11) in 2015, where they initiated the discussion of analyzing how to strengthen the scientific people-to-people exchanges between India and China. For the second author, it was indeed in itself a destiny in his exploration of China from an entirely different perspective. In the past 5 years, this entire intellectual transformative journey for him, which is still continuing, can only be described as profound and mesmerizing.

This article summarizes grass-roots level exchanges between India and China. Such exchanges are not be easy or likely to locate in numerous geo-political discussions or books. It is our equanimity hope that what holds to be self-evident is that the people of these two countries will surely tap into their ancient respective wisdoms and continue to pursue meaningful and sustainable exchanges, irrespective of the current and hopefully temporary geopolitical complexities.

Fig. 6 Glimpses of exchange of Mr. Weiling Cai and Mr. Peinen Chen to IIT Guwahati, India

Fig. 7 Visit of delegation from Shantou University to IIT Guwahati in January 2018. Description from left to right: Mr. Wanshai Shynret (Assistant Registrar and HOS, IIT Guwahati); Dr. Ankit Garg; Dr. Akhil Garg; Dr. Qinhua Wang; Prof Gautam Biswas (then Director, IIT Guwahati); Dr. Xiongbin Peng; Prof. Ravi Punekar (then Dean, Alumni and External Relations office, IIT Guwahati); Prof. Rakhi Chaturvedi (Associate Dean, Alumni and External Relations Office, IIT Guwahati)

Fig. 8 Visit of Shantou University faculties to IIT Bhubaneswar and Interaction with Dr. Hanumantha Rao (left): On the right is picture showing lecture by Dr. Xiongbin Peng, who was invited by Dr. Prasenjit Rath (Assistant Professor, IIT Bhubaneswar)

Fig. 9 Visit of Shantou University faculties to National Institute of Technology (NIT) Meghalaya; Prof Bibhuti Bhusan Biswal (Honorable Director of NIT Meghalaya; second from left), Dr. Xiongbin Peng (center), Dr. Akhil Garg (leftmost), Dr. Ankit Garg (rightmost)

Fig. 10 Author of this article (Prof Feng Da Hsuan) with Prof Bali Deepak (Jawaharlal Nehru University). Picture was taken during his invited visit to University of Macau

Fig. 11 Authors of this article (Dr. Ankit Garg (left) and Prof Da Hsuan Feng (right)). Picture was taken at University of Macau during visit of Dr. Ankit Garg, when he was Assistant Professor at IIT Guwahati in 2015. Prof Da Hsuan Feng was the Special Advisor to President of University of Macau and also Director of Global Affairs

Acknowledgements We are highly grateful to administration of Shantou University, Qingdao University of Technology and Guangxi University for providing assistance in hosting visiting research students from India. We are also grateful to Chinese Scholarship Council for awarding Dr. Gangadhar Reddy and Dr. Ankit Goyal for scholar program at Shantou University. We are indeed thankful to Alumni and External Relations office of IIT Guwahati for efforts in successful visit of two Chinese students under exchange program. We extend deep gratitude for arrangement for visits of Chinese delegation to IIT Bhubaneswar, IIT Guwahati and NIT Meghalaya.

References

1. ISSMGE (2020) https://www.issmge.org/filemanager/article/782/ISSMGE_BULLETIN_2020_JUN_FINAL.pdf
2. Bawa KS, Goodale E, Mtemi W, Chen YF, Barthakur R, Goodale UM, Pandit MK (2020) China and India: toward a sustainable world. Science (New York, NY) 369(6503):515
3. Bawa KS, Goodale E, Mtemi W, Chen YF, Barthakur R, Goodale UM, Liu J, Jiang A, Mammides C, Meegaskumbura M, Pandit MK (2010) China, India, and the environment. Science 327(5972):1457–1459

4. Pandey S (2019) China: a huge draw for Indian students. https://newsin.asia/china-is-a-huge-draw-for-indian-students/
5. ChhapiaH (2018) China gets more Indian students than Britain. https://timesofindia.indiatimes.com/india/china-gets-more-indian-students-than-britain/articleshow/62398336.cms
6. SPARC (2020). https://sparc.iitkgp.ac.in/
7. Council for Indian Institutes of Technology (2020). https://www.iitsystem.ac.in/
8. SPARC (2020). https://sparc.iitkgp.ac.in/apProposal_list.php
9. Department of Mechanical Engineering (2020). https://mech.iitd.ac.in/content/prof-zhou-hongfu
10. Global Initiative of Academic Networks (GIAN) (2019). https://gian.iitkgp.ac.in/ccourses/approvecourses2
11. IIT Bhubaneswar (2019). IIT Bhubaneswar young minds research visit to China international cooperation in less than a year. https://www.iitbbs.ac.in/notice/news_1579261000.pdf
12. ICE (2020). ICE publishing awards 2020. https://www.icevirtuallibrary.com/page/authors/awards-2020
13. Zhu HH, Shi B, Zhang CC (2017) FBG-based monitoring of geohazards: current status and trends. Sensors 17(3):452
14. Kushvaha V (2019) Session on fundamentals of academic writing. https://iitjammu.ac.in/civil_engineering/post/Fundamentals%20of%20academic%20writing